CHEMICAL
REACTION
MECHANISMS

CHEMICAL
REACTION
MECHANISMS

George M. Fleck

Smith College

Holt, Rinehart and Winston, Inc.

New York   Chicago   San Francisco   Atlanta
Dallas   Montreal   Toronto   London   Sydney

Preface

Chemical kinetics deals with chemical reactions that are being watched while things are happening. The emphasis of kinetics is on processes of chemical change, and time is an important experimental variable. The molecular dynamics of the making and breaking of chemical bonds is seen at the microscopic level as a two-way reversible process of reactants and products interconverting. At the same time, at the macroscopic level, the reaction is seen as a one-way process of a nonequilibrium system irreversibly relaxing toward equilibrium.

This work focuses on the experimental reaction-rate curve. Case studies are used to focus attention on representative experimental techniques that yield rate curves; each case study provides open-ended undergraduate kinetics experiments suitable for use in laboratories from the freshman year through senior independent study. In order to interpret the rate curve in chemical terms, a mechanism must be assumed. The mechanism as a unifying intellectual construction, and the rate curve as an experimental fact, are related through the algebra which constitutes a major portion of Chapters 3, 5, 8, and 10.

Illustrative chemical material has been drawn almost entirely from studies of homogeneous reactions in solution. Thus the use of molar concentrations, an emphasis on reactions of complex ions, and the discussion of the influence of pH complement the customary freshman-year topics in solution equilibria. The case studies deal with inorganic and organic experiments in which the reactants can be mixed in beakers at the laboratory bench. This emphasis on solution reactions also recognizes the vital importance of reaction kinetics in the aqueous solution chemistry of experimental biochemistry. One case study is devoted to enzyme kinetics, and there is substantial material in Chapter 10 on the steady-state treatment of enzymatic catalysis. Sophisticated instrumentation may be used if available, but good rate data can be obtained for each of the major case studies in the teaching laboratory with a single-beam spectrophotometer.

v

Most of this book is about first-order and pseudo-first-order reactions. By imposing this limitation, it has been possible to bring considerable understanding of complex reactions within the grasp of the student with a modest mathematical background. Imaginative experimental design can often reduce the mathematical problems of data analysis. There are many experimental ways to achieve pseudo-first-order rate curves for complex reactions, and several of these ways are explored in detail.

The mathematical preparation is assumed to be at a level comparable to that ordinarily achieved in a strong secondary school mathematics sequence. Only minimal use is made of calculus. The notation of the derivative appears in the differential rate equations. The derivative of the exponential is used in solving these differential equations by substitution of a trial solution. An introduction to calculus is, of course, helpful, but my freshman students without calculus have proceeded through this material without serious difficulty, and they have learned well the substantive material of chemical kinetics. Use is also made, late in Chapter 5, of the first and second derivatives to examine a curve for maxima, minima, and inflection points.

The mathematics is mostly the algebra of linear equations. The student should be familiar with the manipulations involved in using exponents and logarithms. Graphical methods are used in treating experimental data. There are numerous mathematical hints and helps at points that have proved troublesome for some students. Explanatory material on determinants and the exponential function appears in appendices. Any student who feels the need to regain mathematical proficiency lost since high school may find the programmed instruction of the following three booklets very helpful: R. J. Flexer and A. S. Flexer, *Programmed Reviews of Mathematics*: Booklet 2, "Linear and Literal Equations"; Booklet 4, "Exponents and Square Roots"; and Booklet 5, "Logarithms." New York: Harper & Row, Publishers, 1967.

The material in this book has been used for a portion of a freshman course, "Structure and Energetics in Chemistry," at Smith College. Sections have also been used in a graduate kinetics course and in a continuing-education seminar for professional chemists. The book is organized in such a way as to be well suited as an introduction to enzyme kinetics for biochemists.

It is a pleasure to acknowledge the direct influence throughout this book of my kinetics teachers: Dr. Robert A. Alberty, Dr. Farrington Daniels, Dr. Edward L. King, Dr. Phillip A. Lyons, and Dr. S. J. Singer.

Northampton, Massachusetts   G.M.F.
July 1970

Contents

Bibliographical Note

References to articles in scientific periodicals are given in an abbreviated form corresponding to the practice of the American Chemical Society. The full name of each journal can be found, together with its accepted abbreviation, in *Access: Key to the Source Literature of the Chemical Sciences*, Columbus, Ohio: Chemical Abstracts Service, 1969. There is a listing, for each periodical, of libraries which are subscribers and an indication of the extent of each library's holdings of back issues. Most college and university libraries subscribe to the common chemical journals and have bound volumes of the back issues on the shelves. In addition, interlibrary loan and photocopying services make available to students and faculty articles in even the very obscure and the very old journals.

CHEMICAL
REACTION
MECHANISMS

1

Concept of Mechanism

When a solution of an EDTA[1] salt in water is mixed with a dilute solution containing hydrated chromium(III) cations, the resulting solution does not appear to the eye to be very much different from the two nearly colorless solutions from which it was prepared. At least there is not much difference at first.

But as time passes and as the solution stands undisturbed in a beaker, all regions of the solution gradually acquire a beautiful violet color. Something is surely happening throughout this solution!

While watching the color intensify without external intervention, most persons think of those natural questions: What is happening? How is it happening? What is the explanation? An answer that can be defended must be based on observation of the beaker and its contents. Can it happen again? Some more of each of the two solutions are mixed again, and the same events occur again. But are they *exactly* the same events? It is hard to be sure, for the design of the experiment places too much reliance on the ability of a human to judge changes in color intensity and changes in time. Some instrumentation—at least a clock and a device for measuring or matching colors—is needed in order to describe the process in terms of numbers.

---

[1] EDTA is ethylene*d*iamine*t*etra*a*cetic acid. The molecular structure of EDTA is found in a detailed discussion of the chromium(III)–EDTA reaction at the beginning of Chapter 2.

1

If data from two experiments can be expressed as numbers, comparison of the experiments *may* be as simple as comparison of these numbers. Chapter 2 deals with many of the facets of designing such experiments. It is possible to express the changing of color by a column of numbers representing color intensity as a function of time, or by a graph of these numbers plotted versus time.

But are these numbers the ones we would expect? This question as asked is not answerable, but there are rather precise answers to the closely related question: Are the numbers the ones expected *if* certain specified chemical reactions were taking place in the solution? We are not totally in the dark about the possible reactions, because we prepared the solutions ourselves from known compounds, and the range of possible reactants is thus restricted. We can make some reasonable guesses—maybe a whole list—of possible reactions that might account for the appearance of violet color. When we get specific and say, in effect: "I wonder if this particular set of chemical reactions could account for the rate at which the color changed," we have proposed a *reaction mechanism*. We then stand in need of a method whereby this mechanism, expressed in terms of chemical equations, can be confronted with the numerical data from the experiments.

The reaction mechanism is a chemical description of the molecular processes occurring throughout the solution. The chemical changes of individual molecules are the elementary reaction steps. Thus the equation

$$A \rightarrow B \tag{1.1}$$

means that a single molecule of compound A reacts to form a single molecule of compound B in an elementary chemical process that does not directly involve any other molecule at the instant of reaction. The equation

$$A + B \rightarrow C \tag{1.2}$$

states that a molecule of A encounters a molecule of B, and that as an immediate result a molecule of C is formed.

Equations (1.1) and (1.2) do not represent complete chemical reaction steps. The principle of microscopic reversibility[2] requires each elementary reaction step to be reversible, and we shall satisfy the requirements of microscopic reversibility by including a reverse arrow in each

---

[2] The principle of microscopic reversibility can be stated as follows: Corresponding to each elementary chemical process there is a reverse process. These two elementary processes together constitute a reversible elementary chemical reaction which can bring about chemical equilibrium between the reactants and products. At equilibrium, the elementary reaction must balance itself, so that the average rate of the forward process is equal to the average rate of the reverse process. The principle of microscopic reversibility is discussed in detail in Chapter 6.

chemical equation. Thus Equations (1.1) and (1.2) should properly be written as

$$A \rightleftharpoons B \qquad \text{(1.1 completed)}$$

$$A + B \rightleftharpoons C \qquad \text{(1.2 completed)}$$

Associated with each forward reaction process and with each reverse process is a reaction-rate constant, symbolized by a lowercase italic k. The larger the value of the rate constant, the faster the chemical change that occurs via a particular elementary reaction process.[3] A particular rate constant will ordinarily have a subscript to identify it and to distinguish it from all other rate constants of the same mechanism. In this book, a rate constant for an elementary process written from left proceeding to right has a positive numerical subscript (thus in Figure 1.1, the rate constant for the forward process is k_1), and the rate constant for the associated elementary process written from right proceeding to left has a negative numerical subscript (thus in Figure 1.1, the rate constant for the reverse process is k_{-1}). Complete characterization of the mechanism of a reaction requires the specification of the chemical equations for all the elementary reactions, and of the numerical values of two rate constants for each chemical equation. The numerical values of the k's are found experimentally by interpreting the quantitative[4] rate data.

How can the mechanism of the chromium(III)–EDTA reaction be found? A *tentative* mechanism is written that takes into account all relevant experimental facts at hand and includes the chemical equations that seem essential in order to get from reactants to products. This proposed mechanism must then be tested against the quantitative facts of chemical reality. Predictions from the mechanism are compared with the observed time course of the color change in the beaker. If there is not good agreement, mechanism and experiment are inconsistent, and something needs to be

---

$$
\begin{array}{ll}
k_1 & k_1 = 6.6 \ \text{sec}^{-1} \\
A \rightleftharpoons B & k_{-1} = 1.8 \times 10^4 \ \text{sec}^{-1} \\
k_{-1} & T = 25.0°\text{C}
\end{array}
$$

FIGURE 1.1. A reaction mechanism.

[3] The relationship between the numerical value of k and the observed rate of the chemical reaction is one of the principal topics discussed in Chapter 3.

[4] The observation that a solution in a beaker is slowly changing from colorless to violet is a *qualitative* observation. When instruments are used so that the data describing the color change are obtained as a table of numbers or as a graph the observation is *quantitative*.

done to improve one or both. If there is agreement, mechanism and experiment are *consistent*. However, it is possible, and indeed probable, that several quite different mechanisms are individually consistent with the same set of experimental data. Additional experiments are then needed to distinguish among the rival mechanisms. The experimenter may discard some mechanisms as inconsistent with these new data. Only those mechanisms consistent with all the data are retained.

BIBLIOGRAPHICAL NOTE

Design and evaluation of kinetics experiments are lucidly discussed by J. F. Bunnett in the chapter, "The Interpretation of Rate Data," in S. L. Friess, E. S. Lewis, and A. Weissberger, eds., *Investigation of Rates and Mechanisms of Reactions*, part I, New York: John Wiley & Sons, Inc. (Interscience Division), 2nd ed., 1961. Particularly recommended for the newcomer to kinetics are pages 178–199 and 271–278. For additional reading of a more general nature about experimental design, see E. B. Wilson, Jr., *An Introduction to Scientific Research*, New York: McGraw-Hill, Inc., 1952.

A clearly written discussion of early experimental and theoretical studies concerning the role of time in chemical reactions is given by E. Farber, *Chymia*, vol. 7, Philadelphia: University of Pennsylvania Press, 1961, p. 135.

Peter Waage and his brother-in-law, Cato Maximilian Guldberg, proposed the law of mass action and laid some of the foundations for chemical kinetics. *The Law of Mass Action, A Centenary Volume: 1864–1964*, Oslo: Universitetsforlaget, 1964, contains a facsimile of the original mass-action paper, biographies of Waage and Guldberg, a history of the discovery, and some rather philosophical current contributions to the theory of chemical kinetics. The law of mass action asserts that the rate of a chemical reaction is proportional to the "active masses" of the reacting chemical species. In this book, the mass action law is modified so as to state that *the rate of an elementary chemical process is proportional to the molar concentrations of each of the reacting chemical species*. In this modified form, the law of mass action leads directly to Equations (3.4) and (3.5) for unimolecular elementary processes, and to Equations (8.5)–(8.7) when a bimolecular elementary process is involved.

2

The Reaction of EDTA and Chromium(III):

A Case Study

The hexaquochromium(III) complex ion

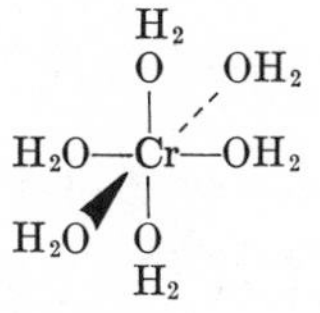

$$
\begin{array}{c}
\mathrm{H_2} \\
\mathrm{O} \quad \mathrm{OH_2} \\
| \quad / \\
\mathrm{H_2O-Cr-OH_2} \\
| \\
\mathrm{H_2O} \quad \mathrm{O} \\
\mathrm{H_2}
\end{array}
$$

forms spontaneously in aqueous solution from chromium(III) salts whenever there is no competition with water from another solute[1] which is better than water at complexing with chromium(III). This complex ion contains six water molecules in an octahedral configuration. These six water molecules are in geometrically (or chemically) equivalent locations at the apices of a regular octahedron.

---

[1] Any species that can occupy one of these six coordination sites around the chromium atom is called a *ligand*. Extensive research projects dealing with rates of replacement of water by various other ligands, and rates of replacement of such ligands by water, have been summarized for Cr(III) in F. Basolo and R. G. Pearson, *Mechanisms of Inorganic Reactions: A Study of Metal Complexes in Solution*, New York: John Wiley & Sons, Inc., 2nd ed., 1967, pp. 196–202.

The reaction in water of the hexaquochromium(III) cation and ethylenediaminetetraacetic acid,

$$
\begin{array}{ccc}
& \overset{\text{H}_2}{\underset{}{\text{C}}} & \overset{\text{H}_2}{\underset{}{\text{C}}} \\
\text{HOOC}\!-\!\text{C} & & \text{C}\!-\!\text{COOH} \\
\diagdown & \text{H } \text{H} & \diagup \\
& \text{N}\!-\!\text{C}\!-\!\text{C}\!-\!\text{N} & \\
\diagup & \text{H } \text{H} & \diagdown \\
\text{HOOC}\!-\!\text{C} & & \text{C}\!-\!\text{COOH} \\
\underset{\text{H}_2}{} & & \underset{\text{H}_2}{}
\end{array}
$$

EDTA or H$_4$Y

yields a complex ion that probably has the schematic structure

$$
\begin{array}{c}
\text{H}_2\text{C}\!-\!\text{N}\!-\!\text{Cr}\!-\!\text{O} \\
\text{H}_2\text{C}\!-\!\text{N}\quad\text{O}
\end{array}
$$

A color change accompanies the formation of the new complex. Depending on experimental conditions, a noticeable color change continues to occur for a period of a few minutes to a whole day. Eventually the solution is violet.

Formation of the chromium(III)–EDTA complex ion involves changes in the chemical environment of the central chromium atom. The octahedral symmetry is lost. Four water molecules are replaced by carboxylate ions of the EDTA molecule, and the remaining two water molecules are replaced by nitrogen atoms of the same EDTA molecule. This EDTA molecule literally wraps itself around the chromium nucleus, changing the environment and perturbing the energy levels of the electrons near the chromium nucleus. It is this change in spacing of energy levels that gives rise to the observed color change.

EDTA is a polyprotic acid. The four carboxylic acid protons can dissociate, and a proton can be bound at each of the nitrogen atoms. Thus the completely deprotonated EDTA molecule, with a net ionic charge of -4, can react with a proton, H$^+$, at any of the binding sites indicated by an arrow:

$$
\begin{array}{ccc}
& \overset{\text{H}_2}{\underset{}{\text{C}}} & \overset{\text{H}_2}{\underset{}{\text{C}}} \\
\rightarrow\!\text{OOC}\!-\!\text{C} & & \text{C}\!-\!\text{COO}\!\leftarrow \\
\diagdown & \text{H } \text{H} & \diagup \\
\rightarrow\!\text{N}\!-\!\text{C}\!-\!\text{C}\!-\!\text{N}\!\leftarrow & \\
\diagup & \text{H } \text{H} & \diagdown \\
\rightarrow\!\text{OOC}\!-\!\text{C} & & \text{C}\!-\!\text{COO}\!\leftarrow \\
\underset{\text{H}_2}{} & & \underset{\text{H}_2}{}
\end{array}
$$

Formation of the chromium(III)–EDTA complex can be thought of as a competition at each EDTA binding site between H$^+$ and Cr(III). The probability of H^+ being bound increases as (H^+) increases,[2] and thus the average number of protons bound to an EDTA molecule depends on the pH of the solution.[3] The differently protonated species might react at different rates, so the observed rate of reaction might depend on pH. Any chemically satisfactory reaction mechanism must deal with the question of how the proton enters into the reaction.

It is difficult to think about the overall reaction in terms of a single elementary reaction step. There is just too much rearrangement of atoms and molecules. A conceptually attractive type of mechanism treats the formation of a chromium(III)–EDTA complex as a stepwise process. By some sequence of simple reaction steps, we can try to visualize how the complicated process of removing six water molecules, attaching the EDTA molecule, and wrapping the EDTA molecule into its most stable configuration can occur around the chromium nucleus. How can such a mechanism be tested experimentally? What is the relationship between this imagined mechanism and the rate of the observed color change?

In the absence of quantitative experimental facts about the reaction, it is difficult to design experiments to test any proposed mechanism. We need some experimental rate data, in the form of numbers, for a particular reacting solution. Then, on the basis of our observations from this *preliminary experiment,* we can proceed to refine the experimental design.

Preliminary Experiment.   Solution I was prepared by dissolving enough disodium ethylenediaminetetraacetate in distilled water to make a 0.1 molar solution. Solution II was prepared by dissolving enough $Cr(NO_3)_3 \cdot 9H_2O$ in distilled water to make a 0.01 molar solution.

Equal volumes of the two solutions were mixed in a beaker. A portion of this mixture was placed in a Pyrex spectrophotometer cell, and the cell was placed in the light beam of a recording spectrophotometer. Figure 2.1 shows plots obtained when the spectrophotometer measured the intensity of light that passed through the solution at various wavelengths of light. An absorbance value of 0.00 means that the same amount of light passed through the solution as would have had the sample been pure water. An absorbance value of 1.00 is obtained when the transmitted light intensity has been decreased tenfold.

An absorbance versus wavelength plot was determined almost immediately after mixing, again after 2 hours, and again after 8 hours. During this time the solution was changing from an almost colorless solution to a solution deep violet in color.

---

[2] The quantity (H^+) is the concentration of H^+, expressed in moles of H^+ per liter of solution. This is a molar concentration.

[3] The quantity pH is approximately equal to $-\log_{10}(H^+)$, and therefore the pH value decreases by one unit for each tenfold increase in (H^+).

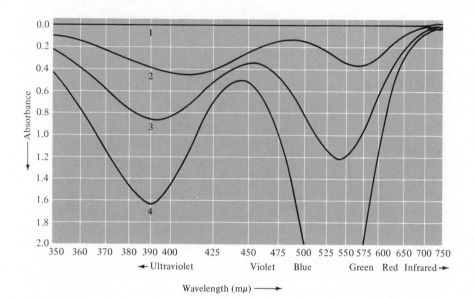

FIGURE 2.1. Spectra of a reacting EDTA–Cr(III) system. Spectra of a reacting EDTA–chromium(III) solution obtained with a Perkin–Elmer Model 350 recording spectrophotometer. The sample was in a quartz cell with a 50-mm optical path length. Spectrum 1: Reference spectrum of distilled water. Spectrum 2: Solution of concentration 0.05 molar Na_2H_2Y, 0.005 molar $Cr(NO_3)_3$, about 5 minutes after mixing. Spectrum 3: Same EDTA–Cr(III) solution, 2 hours after mixing. Spectrum 4: Same EDTA–Cr(III) solution, 8 hours after mixing.

This *preliminary* experiment furnishes enough information about the rate of the reaction and about the accompanying color change to permit the design of an experiment for determining a reaction-rate curve of absorbance versus time. We know from this preliminary experiment about how great the absorbance change will be, about how long the reaction will take, for the particular conditions of EDTA and chromium concentrations, pH, and temperature chosen for the preliminary experiment.[4] Before continuing with discussions of rate experiments, however, we will pause to consider three important topics which bear directly on measurement of these rates of reaction: measurement of absorbance, control of temperature, and the significance of pH.

---

[4] The pH of the solution used for the preliminary experiment was about 4.5. The pH could have been adjusted by the addition of enough base or acid to bring the pH, monitored by a pH meter, to a desired value. The rate of reaction does depend significantly on pH. Investigation and interpretation of the pH dependence would make an interesting student research project. The temperature of the solution during this preliminary experiment was about 23°C.

Measurement of Absorbance. Absorption of visible light by the various chromium(III) complexes occurs when a photon of light is absorbed by an electron in the complex. The electron is promoted to a higher energy level by acquiring the energy of the photon. Such an energy absorption can occur only if the energy-level spacing is equal to the photon energy. When the environment of electrons near the chromium nucleus is altered by changes in the number, nature, or position of bonded ligands, an alteration of electronic energy-level spacings results. Wrapping of EDTA around the chromium nucleus would be expected to perturb the energy levels and change the permitted wavelengths of light that can be absorbed by the complex. The result is a change in the color of the solution containing the reacting complexes.

We are interested in the wavelength of light that is absorbed by the solution, and we are also interested in how much light is absorbed at each wavelength. As chemical reactions proceed in the solution, the amount of light absorbed at each wavelength changes, and this time dependence of light absorbed is directly related to the time changes in concentrations of the reacting chemical species.

A beam of monochromatic light—ideally a beam of parallel rays of light of a single wavelength—will be reduced in intensity while passing through a solution which contains species that absorb photons of that wavelength. The decrease in intensity depends on the number of molecules of the absorbing species in the path of the light beam. If the beam has an intensity I_0 before entering the solution, emerging from the solution with an intensity I, then the Bouguer–Beer law[5] states that, for a solution containing only a single absorbing species X,

$$I = I_0 10^{-l\epsilon_X (X)} \tag{2.1}$$

where l is the optical path length, the distance traveled through the solution by the light beam; (X) is the molar concentration of chemical species X; and ϵ_X is a proportionality constant, called the molar absorptivity of species X.

If there are two different absorbing species in solution, Equation (2.1) is modified to become

$$I = I_0 (10^{-l\epsilon_{X_1}(X_1)})(10^{-l\epsilon_{X_2}(X_2)}) \tag{2.2}$$

---

[5] A derivation and discussion of the Bouguer–Beer law is given in M. G. Mellon, *Analytical Absorption Spectroscopy*, New York: John Wiley & Sons, Inc., 1950, pp. 90–101. This book remains an excellent source of information about spectrophotometry, although continuing rapid advances in instrumentation make it necessary to supplement such a book with the latest information from instrument manufacturers. Alternative derivations of the Bouguer–Beer law are given by H. A. Liebhafsky and H. G. Pfeiffer, *J. Chem. Educ.*, **30**, 450 (1953), and by J. H. Goldstein and R. A. Day, Jr., *J. Chem. Educ.*, **31**, 417 (1954).

Species X_1 and X_2 are assumed to absorb light independently of one another. Equation (2.2) can be written[6] in the equivalent form

$$I = I_0 10^{-l\{\epsilon_{X_1}(X_1)+\epsilon_{X_2}(X_2)\}} \tag{2.3}$$

In general, for a solution containing n different absorbing chemical species, the exponent is given by

$$-l\sum_{i=1}^{n} \epsilon_{X_i}(X_i) \tag{2.4}$$

where the capital sigma, $\sum$, indicates the summation of a series of n terms in which i successively takes on the values $1, 2, \ldots, n$. The absorbing chemical species are $X_1, X_2, \ldots, X_n$. This use of numerical subscripts to differentiate among chemical species is conventional kinetics notation, permitting the straightforward labeling of an arbitrarily large number of different species. Because the notation for X_2 is identical with familiar notation for a dimer containing two molecules of X, confusion could easily result. Ordinarily, the intended meaning of the subscript is obvious.

Inspection of the preceding equations reveals that optical path length l and concentrations (X_i) enter the equations in algebraically equivalent ways. Doubling the concentrations has the same effect as doubling the path length. Figure 2.2 shows schematically the effect on I of changing variables l and (X). If a sample absorbs half the incident light of some

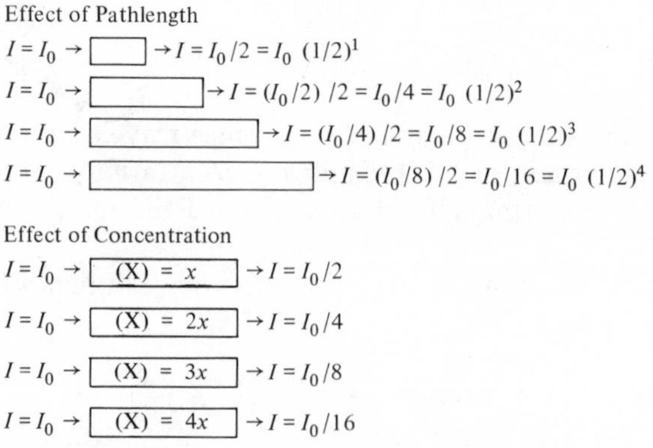

Effect of Pathlength

$I = I_0 \rightarrow \boxed{} \rightarrow I = I_0/2 = I_0\ (1/2)^1$

$I = I_0 \rightarrow \boxed{} \rightarrow I = (I_0/2)\ /2 = I_0/4 = I_0\ (1/2)^2$

$I = I_0 \rightarrow \boxed{} \rightarrow I = (I_0/4)\ /2 = I_0/8 = I_0\ (1/2)^3$

$I = I_0 \rightarrow \boxed{} \rightarrow I = (I_0/8)\ /2 = I_0/16 = I_0\ (1/2)^4$

Effect of Concentration

$I = I_0 \rightarrow \boxed{(X) = x} \rightarrow I = I_0/2$

$I = I_0 \rightarrow \boxed{(X) = 2x} \rightarrow I = I_0/4$

$I = I_0 \rightarrow \boxed{(X) = 3x} \rightarrow I = I_0/8$

$I = I_0 \rightarrow \boxed{(X) = 4x} \rightarrow I = I_0/16$

FIGURE 2.2. Effect of path length and concentration.

[6] To get from Equation (2.2) to Equation (2.3), use was made of the relationship

$$\{10^a\}\{10^b\} = 10^{\{a+b\}}$$

wavelength, a sample with double the concentrations or twice the path-length will reduce the intensity by the factor $(\frac{1}{2})(\frac{1}{2}) = \frac{1}{4}$. A sample that has just the right concentrations and length to halve the light intensity is said to have a length equal to a *half-thickness* for the particular wavelength being used. A sample with an optical path length equal to two half-thicknesses can be thought of as having an imaginary division at its midpoint; light initially having an intensity I_0 will have an intensity of $I_0/2$ as it passes the midpoint having traversed a half-thickness, and its intensity will be halved again as it passes through another half-thickness of sample, finally emerging with an intensity of $I_0/4$.

Quantitative measurements of light absorption in solution are usually reported in terms of either transmittance, T, or absorbance, A. Transmittance is defined by

$$T \equiv \frac{I}{I_0} \tag{2.5}$$

Sometimes the quantity percent transmittance, $\%T \equiv 100T$, is used instead of T. Absorbance is defined by

$$A \equiv \log_{10}\left(\frac{I_0}{I}\right) \tag{2.6}$$

A graphic comparison of the two functions is made in Figure 2.3. There are advantages and limitations to the use of either T or A. Transmittance values are contained within the interval from zero to unity, whereas absorbance values can range from zero to infinity. Transmittance is related

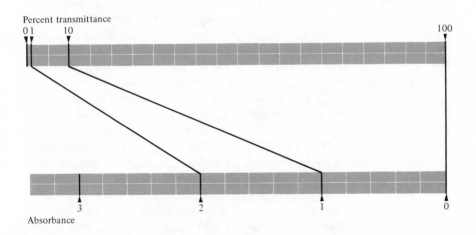

FIGURE 2.3. Comparison of transmittance and absorbance scales.

to concentration in an awkward way. Absorbance is a convenient linear function of concentration, and this fact makes A very useful in determining an experimental reaction rate curve. We shall now investigate how this linear dependence of A on concentration comes about.

Equations (2.1) and (2.6) are combined, taking advantage of two fundamental properties of logarithms:

$$\log_{10} 10^x = x$$

and

$$10^{-x} = \frac{1}{10^x}$$

The result is

$$A = l\epsilon_X(X) \tag{2.7}$$

for a solution containing the single absorbing species X. The general equation relating absorbance to concentrations for a solution with n absorbing chemical species is obtained by combining Equations (2.3), (2.4), and (2.6). The result is

$$A = l \sum_{i=1}^{n} \epsilon_{X_i}(X_i) \tag{2.8}$$

An important chemical research tool is the optical absorption spectrophotometer, an instrument that is used to measure the amount of monochromatic light absorbed by a sample. White light comes from an incandescent tungsten-filament electric light bulb for measurements in the visible portion of the spectrum, and this light is passed through a prism or grating monochromator. Monochromatic light comes from the monochromator as a narrow band of wavelengths distributed around a nominal value shown on the monochromator dial. An absorbance spectrum is obtained by determining the absorbance of the sample at each of many different wavelengths. This spectrum is presented as a plot of A versus wavelength.

The spectra in Figure 2.1 were obtained with a double-beam recording spectrophotometer, an instrument in which wavelength is varied continuously and absorbance plotted versus wavelength automatically. One beam of light passes through the sample solution in a glass or quartz cell, and the other beam passes through air or through a cell containing pure solvent. The intensity of each beam of light is measured with a photomultiplier tube, and the logarithm of the ratio of these two intensities is plotted versus wavelength on a piece of graph paper by a motor-driven recorder pen. The spectra in Figure 2.1 were plotted together with an absorption spectrum of distilled water determined under identical conditions. The water spectrum appears as a nearly straight baseline at about $A = 0$.

Differences in absorbance relative to the water baseline are attributed to absorption of light by solute species. Equation (2.8) justifies an interpretation of the absorbance as being separable into contributions that add together. Thus the absorbance of solvent can be merely subtracted, leaving the absorbance due to solutes. Changes in absorbance with time are attributed to an on-going chemical reaction.

Control of Temperature.   Most reaction rates increase when temperature is increased. Studies[7] of the temperature dependence of the reaction rate of the chromium(III)–EDTA reaction showed that the reaction rate more than doubles when temperature is raised from 25 to 31°C. With this striking dependence of rate on temperature, it is critical for good experimental design to have the temperature of the reacting mixture carefully controlled. A constant-temperature water bath is often used to maintain this close control. A reaction vessel may be immersed in the water bath; or water may be circulated from the bath, around a cell containing the reacting chemicals, and then back into the bath.

Tropical-fish enthusiasts equip their aquarium tanks as automatic constant-temperature baths. They use an electric heater with a thermostat control to keep the water temperature constant at the desired value above room temperature. A water circulator helps keep the temperature uniform throughout the tank. Temperature is monitored with an immersed thermometer. In a small fishbowl with only a small amount of water, the water temperature may keep rising noticeably after the heater shuts off. This overshooting of temperature past the thermostat setting is ordinarily minimal in a large aquarium if there is good circulation of water. In a hot room, it may be necessary to use some sort of cooling arrangement together with the thermoregulated heater to maintain the desired temperature.

A precision laboratory constant-temperature bath is in many ways a refined constant-temperature aquarium. The principal components are a large tank of water, a sensitive thermoregulator, an electric heater, a cooling device, an effective water circulator, and an accurate monitoring thermometer. Several commercially available water baths are designed to maintain temperature constant within the range ±0.005°C. Such high performance requires careful design of the heating, cooling, circulating, and control systems. We shall examine some aspects of the design of an effective constant-temperature bath.

A mercury-in-glass thermoregulator is a favorite temperature-sensing control device. Properly constructed, it is dependable and sensitive; one commonly used model responds to temperature variations as small as 0.001°C. A schematic diagram of a mercury thermoregulator is shown in

[7] R. E. Hamm, *J. Amer. Chem. Soc.*, **75**, 5670 (1953).

Figure 2.4. Mercury metal is liquid in the temperature range of interest, and it expands when temperature increases. Mercury is a good conductor of electricity. Increasing temperature results in expansion of mercury in the large reservoir, and the mercury level rises in the small capillary tube. This thermoregulator is an electric switch that closes and opens an electrical circuit as the mercury level moves up and down at the point of the contact wire in the capillary tube. Typically the height of the contact wire is adjustable so that the thermoregulator can be conveniently set for the desired control temperature. Long-term stability and dependability require that only small electrical currents pass through the thermoregulator, that the mercury be very clean, that the contact wire not react with the mercury, and that the assembly be sealed to exclude the laboratory atmosphere. All four precautions are needed to keep the mercury–wire electrical contact clean so that the making and breaking of the circuit is positive, always occurring at the same mercury level without electrical arcing. A relay is used to switch the high-wattage heater off and on at a signal from the thermoregulator. Use of a relay avoids having high currents pass through the thermoregulator. An electromagnetic mechanical relay can be used. Much smaller control currents, and resultant longer thermoregulator life, can be achieved with an electronic relay using a Thyratron tube.

Heating can be accomplished by a combination of the thermoregulator-controlled electric immersion heater, and an always-on electric immersion heater with adjustable heat input. Cooling can be accomplished by using flowing tap water or some other coolant circulated through a coil of tubing

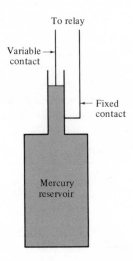

FIGURE 2.4.   Mercury thermoregulator.

immersed in the bath. Thin-wall metal tubing gives good heat transfer. Unless the bath is operating very near room temperature, there will be either heating or cooling by the laboratory environment. The circulating pump or stirrer adds heat to the system. A close balance between the steady heating and cooling processes must be achieved so that the thermoregulated heater can maintain the desired temperature by reasonably frequent off-on switching.

Thorough and rapid circulation of water throughout the bath is essential, and mechanical stirring or pumping is required. Often water is also circulated outside the bath to control temperature in an external reaction cell.

The entire justification of this elaborate temperature control system is the maintenance of constant-temperature conditions within the reacting solution. A monitoring thermometer either within or very near the reacting solution can serve to measure the average temperature of the solution and also to reveal the size of the inevitable fluctuations about this average value. To observe fluctuations, it is necessary to have a thermometer with a sufficiently expanded scale. Since an occasional thermometer may be grossly in error, and many thermometers give readings that deviate from standard temperature values by several scale divisions, it is good practice to use a calibrated thermometer. A simple check to reveal gross inaccuracy can be made by taking the temperature of a distilled water–ice mixture in a Thermos bottle. This temperature should be very close to 0°C. However, an expanded-scale thermometer may not have the ice-point temperature on scale. If the value of the temperature must be known with certainty, it is wise to calibrate the monitoring thermometer with a thermometer of the same range which has been calibrated by the National Bureau of Standards, Washington, D.C. This calibration versus an N.B.S. standard can be performed by placing the two thermometers side by side in an operating constant-temperature bath, waiting until thermal equilibrium has been achieved before comparing readings. Sometimes the mercury column sticks within the fine capillary of the thermometer, and a gentle tap before reading may be needed.

Significance of pH. In water solution, EDTA can exist as five different chemical species, denoted as H_4Y, H_3Y, H_2Y, HY, and Y, the differences arising in the total number of bound protons per molecule. To simplify notation, no ionic charges are indicated on these species. We can expect that each species will react at a different rate. Since the relative number of each of these species depends on the pH value of the solution, we would predict that the observed rates of reaction will depend on pH. In fact, both spectra and the reaction rate do depend on pH. Interpretation of this pH dependence might well give a chemist a great deal of insight into the mechanism of the chromium(III)–EDTA reaction.

The simplest possibility would be that only one ionic species reacts. A more complicated situation sould involve additional species reacting at different rates. There are, in fact, more individual species than is apparent in the listing of $H_4Y, \ldots, Y$. Detailed indication of ionic charges can be quite involved. For example, the species H_4Y has an overall net charge of zero, but there are several ways to arrange the four protons. A proton bound at a nitrogen atom contributes a positive charge in that region, while the absence of a proton at a carboxylate group leaves a negative charge there. So H_4Y can exist as H_4Y^0, $H_4Y^\pm$, and $H_4Y^{\pm\;\pm}$. The distribution diagram in Figure 2.5 shows how the total EDTA concentration, $[Y]$, is distributed among the concentrations of the variously protonated forms as a function of pH. If all the EDTA were present as H_2Y, the distribution fraction $(H_2Y)/[Y]$ would be equal to unity and all the other distribution fractions would be equal to zero. The distribution diagram in Figure 2.5 was calculated from data reported by Schwarzenbach and Ackermann.[8] Information such as that contained in this distribution diagram can be combined with data from rate experiments at various values of pH to construct a detailed mechanistic picture of exactly how this reaction takes place.

No one yet has reported such a detailed analysis of rate data for the chromium(III)–EDTA reaction.

Experimental Rate Curve

We now have enough information to design a good experiment for the determination of a reaction-rate curve for the chromium(III)–EDTA reaction. Certain details depend on the equipment available in the laboratory and on the importance attached to the results. Two different experiments will be described. They yield essentially the same information, but they make use of different types of thermostatting and spectrophotometric equipment.

Sampling Method. In this experiment, a large volume of reacting solution is kept at constant temperature in a flask immersed in a water bath. At convenient time intervals, samples are withdrawn and transferred to a spectrophotometer cell. The absorbance of the sample is measured with a spectrophotometer, and both time and absorbance value are recorded for each measurement. Can the sample then be poured back into the large flask? Is temperature control of the sample necessary while it is in the spectrophotometer?

An appreciable amount of time is required for thermal equilibration of a large volume of reactant with the constant-temperature bath, and

---

[8] G. Schwarzenbach and H. Ackermann, *Helv. Chim. Acta*, **30**, 1798 (1947).

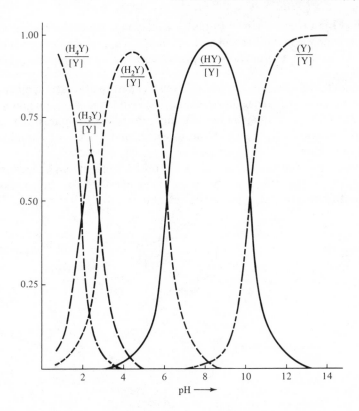

FIGURE 2.5.   Distribution of variously protonated species of EDTA.

SOURCE: G. M. Fleck, *Equilibria in Solution,* New York: Holt, Rinehart and Winston, Inc., 1966, p. 186. Reprinted by permission.

so if the reactants are mixed and *then* the mixture placed in the thermostat, data from the early portion of the reaction will be suspect. That early data will have been taken at an unknown and changing temperature. A better method is to equilibrate two solutions in the water bath and mix these two solutions to initiate the reaction.

How many measurements should be made? It is good practice to get data throughout at least two half-lives of the reaction. If the infinite-time method or half-life method[9] is to be used for evaluation of the rate constant, the limiting value of absorbance must be measured after the solution has reacted for many half-lives. Enough pairs of time–absorbance numbers should be obtained so that the points plotted on a piece of graph paper can

[9] See pages 36 and 44.

be joined by a smooth curve to give a continuous time–absorbance graph. If too few points have been obtained from the experiment, it may not be clear how to draw the portions of the curve between the points.

Continuous Method. A rather different experimental arrangement can be used to obtain the reaction-rate curve as a continuous ink line on a piece of strip-chart recorder paper. Little if any effort by the chemist is required *after* the reaction has begun. More advance preparation of equipment is needed, however.

Many spectrophotometers can be adapted to receive a thermostatted cell so that temperature can be maintained constant for long periods of time. If this can be done, the reacting solution can be left in place in the

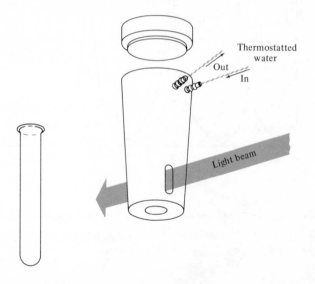

FIGURE 2.6. A thermostatted cell holder for a Spectronic 20 spectrophotometer. A glass test tube is the spectrophotometer cell, fitting into the long hole along the axis of the aluminum cell holder. The light beam of the spectrophotometer passes through a slit in the cell holder, through the test tube with the reacting solution inside, and finally out through a second slit in the holder. Thermostatted water is circulated through hollow passages inside the aluminum cell holder, maintaining the metal holder, the test tube, and the reacting solution at the water temperature. Two tubulatures receive rubber tubing which carries thermostatted water to and from the constant-temperature water bath. The metal cover serves a dual purpose. The cover adds sufficient weight to the test tube so that a portion of the test tube protruding from the bottom of the cell holder can depress a shutter lever in the Spectronic instrument. The cover also excludes light, although it has been found in practice that a black cloth over the assembly is needed if there is bright illumination in the laboratory room.

spectrophotometer throughout the reaction. Many spectrophotometers, including some of the less-expensive models, have provision for recording absorbance (or transmittance) continuously as a function of time. If the instrument has a built-in recorder, it is only necessary to be able to have the chart paper move at a constant rate without the wavelength being changed. Some spectrophotometers without recorders have connections

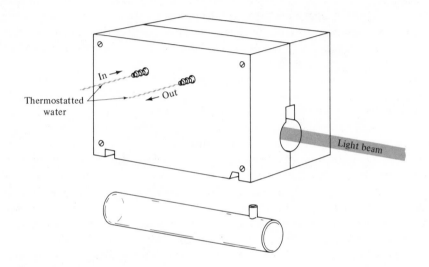

FIGURE 2.7. A thermostatted cell holder for a Perkin-Elmer Model 350 spectrophotometer. This hollowed aluminum block is designed to hold a cylindrical glass cell and to rest on two existing support rods in the spectrophotometer cell compartment. The block was made in two pieces to facilitate the machining of internal chambers for water circulation. The left- and right-hand pieces are fastened together with four machine screws, a gasket cut from thin rubber sheeting forming a watertight seal between the sections. A rectangular cutout is required to accommodate the upright filling tube of the glass cell. Two tubulatures receive rubber tubing which carries thermostatted water to and from the constant-temperature water bath. There is less problem with reflection of light if the block is given a flat-black finish. A thermostatted cell block to fit a Beckman DU spectrophotometer long-cell compartment can be made in the same way, with only some of the dimensions altered. For both the Perkin–Elmer and the Beckman instruments, a method is required to exclude room light from the cell compartment, while permitting the rubber tubing to lead from the cell holder to the water bath. A loosely fitting plywood lid is satisfactory, but such an arrangement should be supplemented with a black cloth over the cell compartment. It is helpful, if feasible, to reduce the intensity of room illumination directly over the spectrophotometer. A block to fit the Cary Model 14 spectrophotometer can be made with an integral cell compartment lid, and the rubber tubing can then be connected from above, eliminating problems with light entry.

for attaching a separate recorder. In any case, the rate at which the chart paper moves through the recorder must be known.

Once again, greater confidence in the early data will result if the reaction is initiated by mixing two solutions, each already at the reaction temperature.

Some Results. We shall now interpret the experimental reaction-rate curve. Details of how such interpretation might proceed, with descriptions in terms of chemical reaction mechanisms, constitute the subject matter of most of the remaining pages of this book. There are a few general comments that can be made here.

If it is assumed that the color change is really the result of just a single elementary chemical reaction in which some species A converts to a different species B, an algebraic equation can be derived for the change of absorbance with time. The derived equation has the same form as the experimental curve obtained at pH 4.5 with concentration of EDTA large compared to the concentration of chromium(III) species. But this explanation, although fitting the facts obtained from a single experiment, is not complete and may even be misleading. There needs to be at least one step involving the encounter of two reactants. At some values of pH, the curve is more complicated, and the shape of the rate curve is that predicted if there were two elementary reactions in the mechanism. The dependence of rates of reaction on initial concentrations of the various species in solution can give valuable information about these elementary reactions. The hydrated proton, H^+, is involved in the reactions, maybe directly as a reactant, or maybe indirectly by determining relative proportions of EDTA species. And there is the hope that somehow some information about the actual wrapping-around process can be extracted from this rate data.

It seems possible to study this reaction in great detail without meeting experimental rate curves other than first-order curves such as are discussed in Chapter 3 and sums of first-order curves such as are discussed in Chapter 5. For these first-order curves, each experiment yields one or more *macroscopic first-order rate constants*. The primary focus in designing experiments for studying the chromium(III)–EDTA reaction is on the effect that variation of initial concentrations has on the numerical values of these experimental macroscopic rate constants. The interpretation of concentration-dependent macroscopic first-order rate constants is discussed throughout Chapter 8. To keep the number of variables reasonably small, it seems prudent to perform all these experiments at the same temperature. It is also wise to study the effect of concentration on rate by changing just one concentration value at a time, keeping all other concentrations constant within a series of rate experiments.

SUGGESTIONS FOR EXPERIMENTAL INVESTIGATION

A detailed study of the influence of each of the many experimental variables on the rate of the chromium(III)–EDTA reaction, with enough duplicate experiments to verify reproducibility, might turn out to be a project of thesis proportions for a student working independently. Or with just one experimental setup, a class can obtain quite a bit of experimental data during a semester if a different student or group of students performs a rate experiment at a different set of concentrations each laboratory period. Some straightforward and important experimental questions can be answered by a series of rate experiments: Does the rate depend on initial concentration of EDTA? Does the rate depend on initial concentration of chromium(III) salt? Does the rate depend on initial concentration of H^+?

When answers have been obtained to these questions, the time has come to propose mechanisms to account for the accumulated experimental facts. The predictions of the proposed mechanisms can be made quantitative, using methods developed in the succeeding chapters. Then experiments can be designed with care to pick the best concentration ranges for critical tests of the proposals. At this point the chemist is in a position to design experiments on the basis of his own previous data. The aim is to obtain data that will allow some of the mechanisms to be rejected as inconsistent with the experiments, and at the same time to obtain data for the evaluation of the individual *microscopic rate constants* of the mechanism or mechanisms consistent with all the data. An assignment, which could serve as a review of much of the material in this book as well as a laboratory report of the class results, would be a research proposal that would propose a mechanism consistent with the class results and would give the experimental details, complete with suggested concentration ranges, for the experiments to test the mechanism.

Suppose that these experiments showed that the rate of this reaction increases with increasing concentration of H^+, and therefore decreases with increasing concentration of OH^-, for a particular set of reacting conditions. We would describe this observed phenomenon by saying that H^+ is a positive catalyst of the reaction and that OH^- is a negative catalyst of the reaction, under the particular reaction conditions used. This description is a phenomenological, macroscopic description. It tells *what* happens, but it does not attempt to tell *why* or by what mechanism the reaction rate is changed; it tells what is observed happening to the whole system, but it is not concerned with the interactions of individual molecules. One of the insistent demands upon chemical kineticists is that they elucidate the molecular mechanisms of the catalysis of chemical reactions. Progress has been made in understanding catalysis when microscopic descriptions (consistent with the quantitative experimental data from rate studies) can be made in which the catalyst molecule has been written as a reactant molecule in chemical reactions of the kinetic mechanism.

Hedrick[10],[11] has described undergraduate kinetics experiments based on the chromium(III)–EDTA reaction, including an experiment that investigates the catalysis of the reaction in carbonate solutions. A student should be ready to modify details of these experiments to suit the equipment available to him and to increase the accuracy and reliability of the data. For instance, in any critical test of mechanism, thermostatting would seem to be required unless the experiment is being performed at room temperature and the room temperature actually stays constant during the period of the experiment. The cell compartments of most spectrophotometers are appreciably warmer than room temperature, and the cell-compartment temperature may continue to increase during a kinetic run.

The primary purpose of such detailed chemical investigations is the assembling of experimental evidence to support a theory about the molecular chemistry that takes place in these solutions. During the course of this research, several different mechanisms may be proposed and then subsequently modified or discarded. Most tentative mechanisms will probably have features which suggest additional experiments. Throughout the research, there is a continuing and intimate interinvolvement of experiment and theory.

Glycine–Chromium(III) Reactions. The rate of the reaction of the hexaquochromium(III) cation with three molecules of the amino acid glycine,

$$\begin{array}{c} \text{H} \\ ^+\text{H}_3\text{N}-\text{C}-\text{COO}^- \\ \text{H} \end{array}$$

has been followed spectrophotometrically[12] in the visible region of the spectrum. The reaction probably involves a stepwise addition of the glycine ligands. With glycine concentrations about 100 times greater than the concentration of chromium(III) species, plots of $\log_{10} \{A_\infty - A_t\}$ were found to be linear.[13] The researchers cite evidence for the existence of a fast relaxation and of one or more slow relaxations, but their measurements were of a first-order relaxation slow enough so that reactants can be mixed

[10] C. E. Hedrick, *J. Chem. Educ.*, **42,** 479 (1965).

[11] S. R. M. Agger and C. E. Hedrick, *J. Chem. Educ.*, **43,** 541 (1966).

[12] D. Banerjea and S. D. Chandhuri, *J. Inorg. Nucl. Chem.*, **30,** 871 (1968).

[13] See "Infinite-Time Method" (p. 36) and "Constant-Concentration Approximation" (p. 120).

in a beaker, poured into a spectrophotometer cell, and then absorbance measured as a function of time—for times of the order of 1 hour or so. Can experiments be designed to provide information about these other relaxations? What is the minimum mechanism that takes into account all the experimental evidence cited by the authors? What are the possibilities of extending the work of these chemists to the evaluation of several of the microscopic rate constants of a mechanism?

Problems

2.1. Calculate the absorbance that corresponds to each of the following values of transmittance: $T = 1.00, 0.10, 0.01, 0.97, 0.96, 0.03, 0.04$. It is assumed that these transmittance values were read from a spectrophotometer scale, linear between 0.00 and 1.00, on which numerical values could be estimated within about 1 percent of full scale. Indicate in your answers the uncertainty in each absorbance value which results from the uncertainty of ± 0.005 inherent in reading the transmittance value.

2.2. Calculate the value of (H^+) that corresponds to each of the following pH values: pH $= 14.00, 7.00, 2.00, 2.01, 1.99, 2.50$.

2.3. The absorbance of a certain solution is 20 when contained in a cell with an optical path length of 100 mm. The spectrophotometer available has an absorbance range of 0 to 2, so measurements cannot be made on such a long cell. Find the absorbance of this solution in cells of the following path lengths: $l = 10.0$ mm, 1.00 mm, 0.10 mm. There are commercially available cells with path lengths throughout this range. Filling is not difficult, even with the shortest path length.

BIBLIOGRAPHICAL NOTE

A helpful comparison of various pictorial representations of the three-dimensional structure of octahedral complexes appears in chapter 1 of F. Basolo and R. G. Pearson, *Mechanisms of Inorganic Reactions: A Study of Metal Complexes in Solution*, New York: John Wiley & Sons, Inc., 2nd ed., 1967. A discussion of the kinetics and mechanisms of the closely related cobalt(III)–EDTA reactions, with emphasis on the geometrical aspects of the reactions, appears later in the same book, on pages 321 and 341–343.

The rates of the chromium(III)–EDTA reaction have been studied through the pH range 1.65 to 5.86, and through the temperature range 25.2 to 42.5°C by R. E. Hamm, *J. Amer. Chem. Soc.*, **75**, 5670 (1953). Comparison is made in this paper between the reactions of chromium(III) with the oxalate ion and with EDTA. A three-step mechanism is proposed as consistent with the rate experiments for the EDTA reaction.

Acid–base equilibria in this reaction may be ignored if all experiments are performed in solutions of the same pH. However, much about the mechanism can be learned from analysis of data obtained from experiments at various values of pH. A student needing background for dealing with the varied aspects of the multiple equilibria in these solutions is referred to G. M. Fleck, *Equilibria in Solution,* New York: Holt, Rinehart and Winston, Inc., 1966, particularly to a section on measurement of pH, pages 41–44, and to a case study on EDTA, pages 182–188.

Detailed kinetic analysis of reactions involving the wrapping and unwrapping of a molecule such as EDTA around a metal ion is both fascinating and complicated. One such study [D. W. Margerum, D. L. Janes, and H. M. Rosen, *J. Amer. Chem. Soc.,* **87,** 4463 (1965)] of the stepwise nature of the unwrapping and transfer of an EDTA ligand from nickel(II) to copper(II) makes challenging and rewarding reading for the student who has completed most of this textbook.

A valuable guide to spectrophotometric measurements, dealing mainly with how to obtain dependable absorption spectra within the wavelength range 200 to 800 mμ, is J. R. Edisbury, *Practical Hints on Absorption Spectrometry (Ultra-violet and Visible),* New York: Plenum Publishing Corporation, 1967.

An authoritative discussion of constant temperature baths is given by E. D. West in I. M. Kolthoff and P. J. Elving, eds., *Treatise on Analytical Chemistry,* part I, vol. 8, New York: John Wiley & Sons, Inc. (Interscience Division), 1968, chap. 93. Important precautions that should be observed to obtain the highest accuracy from thermometers are described in National Bureau of Standards Monograph 90, *Calibration of Liquid in Glass Thermometers,* Washington, D.C.: U.S. Government Printing Office, 1965. L. Mészáros [*J. Chem. Educ.,* **47,** 244 (1970)] has described an ultrathermostat in which uniform heating is achieved by passing an electric current through an electrolyte solution which is also the thermostatting medium. An excellent article on precise control of temperature in chemical experimentation is M. Van Swaay, *J. Chem. Educ.,* **46,** A515, A565 (1969).

The physical chemist E. B. Wilson, Jr., has written an excellent guidebook to the design of chemical experiments and research projects, *An Introduction to Scientific Research,* New York: McGraw-Hill, Inc., 1952. This book has much to offer both the novice and the experienced chemist.

Accurate measurement of time is important in every rate experiment. For time intervals of the order of minutes and hours, reliable time measurement can be made with a clock, a timer, or a strip-chart recorder driven by a synchronous electric motor. Such a time measurement depends on the frequency accuracy of the nominal 60-hertz alternating current supplied by the electric utility, and this frequency in turn depends on comparison with signals from the National Bureau of Standards. For information about the practical definition of the U.S. standard time interval, its relationship to international standards, and about the way this standard is disseminated, see A. V. Astin, *Sci. Amer.*, **218**(6), 50 (1968); and J. M. Richardson and J. F. Brockman, *Phys. Teacher*, **4**, 247 (1966). A careful study [R. S. Craig, C. B. Satterthwaite, and W. E. Wallace, *Anal. Chem.*, **20**, 555 (1948)] of the errors encountered in measuring time intervals of a few minutes, relying on the commercial 60-hertz power available in Pittsburgh in 1947, showed maximum deviations of less than 0.2 percent and probable errors of about 0.05 percent, owing to fluctuations in the alternating-current frequency.

3

Rate of a Chemical Reaction

Suppose that there exist in solution two different compounds, B and C, which spontaneously interconvert. As is the case with all chemical conversions, this reaction is not instantaneous. Time is required to get from a state far from equilibrium to another state close to equilibrium. The simplest possible molecular mechanism describing the time course of this chemical interconversion can be represented by the two equations

$$B \xrightarrow{k_1} C \qquad\qquad (3.1)$$

$$B \xleftarrow[k_{-1}]{} C \qquad\qquad (3.2)$$

A more compact notation is usually used when writing such a mechanism, combining the forward and reverse processes into a single chemical equation to give the elementary reaction

$$B \underset{k_{-1}}{\overset{k_1}{\rightleftharpoons}} C \qquad\qquad (3.3)$$

Equation (3.3) has exactly the same meaning as Equations (3.1) and (3.2) considered together. What is that meaning? Equation (3.3), considered as the chemical statement of the complete mechanism of the reaction, means:

1. There exist no kinetically significant chemical species in solution other than compound B and compound C.

2. The rate[1] of production of compound C from compound B via Process (3.1) is proportional to the number of molecules of B in solution, and therefore is proportional to the molar[2] concentration (B). Written in terms of the time derivative of the molar concentration (C), the statement becomes

$$\left(\frac{d(\mathrm{C})}{dt}\right)_{Process\ (3.1)} = k_1(\mathrm{B}) \tag{3.4}$$

The rate of consumption of C by production of B via Process (3.2) is

$$\left(-\frac{d(\mathrm{C})}{dt}\right)_{Process\ (3.2)} = k_{-1}(\mathrm{C}) \tag{3.5}$$

The symbol t denotes time. The quantities (B) and (C) are instantaneous

---

[1] The rate of production of compound C is the rate of change of the concentration of C with time, which is the time derivative of the concentration. Thus

$$\left(\begin{array}{l}\text{rate of change of (C)}\\\text{with respect to time}\end{array}\right) = \left(\frac{d(\mathrm{C})}{dt}\right) = \left(\begin{array}{l}\text{time derivative}\\\text{of (C)}\end{array}\right)$$

The time derivative of the concentration at a particular time t is the slope of a graph of concentration versus time at that particular time. If the graph were a straight line, the numerical value of the slope would have been the same at every point on the graph. The slope could have been calculated by the equation

$$\text{slope of straight line} = \frac{\delta(\mathrm{C})}{\delta t}$$

where $\delta(\mathrm{C})$ is the change in the value of (C) during the time interval of length δt. For the straight-line graph, any time interval and corresponding $\delta(\mathrm{C})$ will give the same slope. However, the concentration–time graph is seldom a straight line, and, in fact, the slope is usually different at each value of t. The value of the slope at a point on the graph is then defined by the more general equation, expressed in terms of the limit in which the size of the time interval δt becomes very small:

$$\text{slope of line} = \lim_{\delta t \to 0} \frac{\delta(\mathrm{C})}{\delta t} = \frac{d(\mathrm{C})}{dt}$$

[2] Rate of reaction, expressed in number of molecules per unit volume transformed per unit of time, is proportional to the number of molecules of reactant in that unit of volume. The molar concentration scale is especially convenient for writing the differential rate equations, because molarity is proportional to the number of molecules per liter. Other familiar concentration scales, such as the molal scale and the mole fraction scale, are not defined in terms of volume and consequently are not used for rate equations.

molar concentrations of the chemical species B and C in solution. Both concentrations and both derivatives are time-dependent variables whose numerical values change continuously as the reacting system evolves toward a state of chemical equilibrium. The rate of change of a concentration is negative if that chemical species is disappearing, positive if the species is appearing. The proportionality constants k_1 and k_{-1} are the rate constants of the mechanism. These rate constants have numerical values[3] which are independent of time but which are in general functions of such experimental variables as temperature and pressure. In particular, most rate constants depend strikingly on temperature, with the dependence of k on temperature T usually of the form[4]

$$k = Z \exp(-\Delta E_{\text{activation}}/RT) \tag{3.6}$$

where the preexponential coefficient Z is approximately independent of temperature. Temperature is expressed on the absolute Kelvin scale.[5] The quantity R is the gas constant, equal to 1.987 calories·degree^{-1}·mole^{-1}. The quantity $\Delta E_{\text{activation}}$ is called the activation energy for the process. For many chemical reactions in solution, $\Delta E_{\text{activation}}$ has a value such that the exponential doubles in value when temperature is increased by only about 10 degrees. Since rather small changes in temperature can have a large effect on reaction rate, it is essential that temperature be kept as nearly constant as possible during a kinetics experiment. Rather elaborate thermostatting equipment is often used to keep solution temperatures constant within a range of $\pm 0.01°$C.

3. The overall net rate of change of the concentration of C is found by superimposing the opposing processes of formation and consumption of C. The net rate of the reaction is the sum of the rates for the separate processes. This overall rate of change is denoted by the derivative without

---

[3] The time derivative in Equation (3.4) is expressed in the units moles·liter^{-1}·second^{-1}. Since both sides of the equation must have the same units, and the concentration (B) has the units moles·liter^{-1}, the rate constant must have the units second^{-1}. This k is a unimolecular rate constant; bimolecular rate constants with the units seconds^{-1} [moles/liter]$^{-1}$ are introduced in Chapters 7 and 8.

[4] Equation (3.6) is discussed in Chapter 12, with particular emphasis on the chemical significance of the activation energy. Table 12.1 illustrates, for several different numerical values of the activation energy, the variation in the value of the exponential factor arising from a 10-degree change in T near room temperature. The activation energy is interpreted as the energy required to form an activated molecule from a reactant molecule. The activated species then may convert into a product molecule. The notation $y = \exp(x)$ is equivalent to $y = e^x$.

[5] The two centigrade temperature scales are related by an additive constant. To obtain the temperature value on the Kelvin scale, °K, add 273.15 to the customary temperature on the Celsius scale, °C.

a qualifying subscript. Thus

$$\frac{d(C)}{dt} = \left(\frac{d(C)}{dt}\right)_{Process\ (3.1)} + \left(\frac{d(C)}{dt}\right)_{Process\ (3.2)}$$

$$\frac{d(C)}{dt} = k_1(B) - k_{-1}(C) \tag{3.7}$$

4. Likewise, the overall net rate of change of the concentration of B is given by

$$\frac{d(B)}{dt} = \left(\frac{d(B)}{dt}\right)_{Process\ (3.1)} + \left(\frac{d(B)}{dt}\right)_{Process\ (3.2)}$$

$$\frac{d(B)}{dt} = -k_1(B) + k_{-1}(C) \tag{3.8}$$

5. In Process (3.1) and also in Process (3.2), the chemical transformation is considered to occur as if there were no direct interaction of the reacting molecule with any particular other molecules at the time when reaction occurs. Only one molecule reacts in each process, and so each elementary process is said to be *unimolecular*, to have a *molecularity* of unity. The number of molecules involved as reactants in an elementary process is the molecularity of that process. Mechanism (3.3) specifically excludes the type of reaction in which chemical transformation occurs as the immediate and direct result of a molecular collision between two reactants; such a transformation would be called a bimolecular process.[6]

6. Equations (3.7) and (3.8) show that the rate of change of each concentration is given by a sum of terms, each term containing just one concentration factor. A term with one concentration factor is a *first-order* term. Since all the terms in the right-hand members of Equations (3.7) and (3.8) are first-order terms, these equations are called *first-order rate equations*.[7]

Integrated Rate Equations

We are trying to make possible a meaningful and informative confrontation of the results of a rate experiment with the predictions of a chemical model—the chemical reaction mechanism. Experimental kinetic

---

[6] A bimolecular process is examined at the beginning of Chapter 8.

[7] This is the conventional usage among chemical kineticists. Care must be taken to avoid confusion with the mathematician's definition of the *order of a differential equation*, by which is meant the order of the highest derivative in the equation. In this book, all the rate equations are first-order differential equations in the sense that the order of the derivative is unity.

data are usually obtained as plots of some property of the reacting system versus time. For example, the experimental rate curve discussed in Chapter 2 was a plot of optical absorbance versus time. To confront a chemical model with these experimental data, it is first necessary to formulate a mathematical model, and then to derive an equation for the time–property plots predicted by the mathematical model. These predicted rate plots are compared with the experimental rate plots.

Equations (3.7) and (3.8) together constitute the mathematical model for the reaction mechanism. The quantitative predictions for the mechanism are deduced from this mathematical model. Equations (3.7) and (3.8) are simultaneous equations which together describe the evolution of the concentrations (B) and (C). Because each contains a derivative, each is called a differential equation. Associated with these differential equations are some ordinary algebraic equations, relating the concentrations to time, which can be obtained by the process of integrating the differential equations. Integration of some differential equations is difficult, and the integration of many differential equations is probably impossible. We are fortunate, because solutions are known for our equations. The integrated equations that we seek can be found by a substitution method. Success of the method depends on the choice of an appropriate trial function. Happily, mathematicians before us have discovered the trial function that we require.

Trial functions[8] of the form

$$(B) = be^{-mt} \tag{3.9}$$

$$(C) = ce^{-mt} \tag{3.10}$$

will be substituted into Equations (3.7) and (3.8). These trial functions will be judged to be satisfactory solutions if this substitution yields a set of equations that is valid for all values of t from zero to infinity and for arbitrary[9] values of b and c. To perform this substitution, we need to know the functional form of the two derivatives

$$\frac{d(B)}{dt} \quad \text{and} \quad \frac{d(C)}{dt}$$

The exponential function has the convenient property (for our derivation, this is an essential property) that its derivative is proportional to the orig-

---

[8] The exponential function e^{-mt} and its time derivative are discussed in Appendix II.

[9] The phrase "arbitrary values of b and c" means "any arbitrarily chosen values of b and c," with the possible restriction of either or both b and c to some range of numerical values.

inal function. In particular, the derivative of Equation (3.9) is

$$\frac{d(B)}{dt} = -mbe^{-mt} \qquad (3.11)$$

and the derivative of Equation (3.10) is

$$\frac{d(C)}{dt} = -mce^{-mt} \qquad (3.12)$$

Substitution of Equations (3.9), (3.10), (3.11), and (3.12) into Equations (3.7) and (3.8) yields

$$-mbe^{-mt} = -k_1be^{-mt} + k_{-1}ce^{-mt} \qquad (3.13)$$

$$-mce^{-mt} = k_1be^{-mt} - k_{-1}ce^{-mt} \qquad (3.14)$$

Division by the common factor e^{-mt} leaves two algebraic equations in which the variable t is absent:

$$-mb = -k_1b + k_{-1}c \qquad (3.15)$$

$$-mc = k_1b - k_{-1}c \qquad (3.16)$$

These last two equations are independent of the instantaneous value of t, because neither t nor any time-dependent quantity appears in the equations. Rather than describing the time course of a particular reacting system, these two simultaneous equations state some algebraic conditions that must be satisfied for the many conceivable reactions in which the initial values of (B) and (C) are varied over a wide range. The two rate constants k_1 and k_{-1} are considered to be independent of the concentrations (B) and (C), and so the k's are not only independent of concentration and time in any particular experiment but also are independent of the initial concentrations of B and C in a series of experiments performed at the same temperature. The two coefficients b and c are functions of the initial concentrations[10] of B and C, and thus b and c can assume values throughout a wide range while k_1 and k_{-1} are constant. Equations (3.15) and (3.16) must be simultaneously valid for arbitrary values of b and c. The question to be answered next is: What is the necessary relationship between m and the two k's for the case of constant k's and independently variable b and c?

---

[10] See page 97 for the explicit relationship between the preexponential coefficients and the values of the initial concentrations for a similar mechanism.

We proceed to eliminate both b and c from Equations (3.15) and (3.16). Solving both equations for b gives

$$b = \frac{[k_{-1} - m]c}{k_1} \tag{3.17}$$

$$b = \frac{ck_{-1}}{k_1 - m} \tag{3.18}$$

Eliminating b between Equations (3.17) and (3.18) (that is, setting the right-hand members of the two equations equal to one another) results in

$$\frac{[k_{-1} - m]c}{k_1} = \frac{k_{-1}c}{k_1 - m} \tag{3.19}$$

Division of both sides of Equation (3.19) by c, followed by slight rearrangement, gives

$$[k_{-1} - m][k_1 - m] = k_{-1}k_1$$

or

$$k_{-1}k_1 - mk_1 - mk_{-1} + m^2 = k_{-1}k_1$$

or, finally,

$$m[k_1 + k_{-1}] = m^2 \tag{3.20}$$

Two distinct values of m satisfy Equation (3.20). These two roots of the equation will be labeled m_0 and m_1 and will be called the two *macroscopic rate constants* of the reaction. The roots are

$$m = m_0 = 0$$

$$m = m_1 = k_1 + k_{-1} \tag{3.21}$$

The numerical value of the nonzero macroscopic rate constant m_1 can be evaluated directly from an experimental rate curve by methods that are described in detail in the next section. A macroscopic rate constant is a phenomenological quantity that characterizes the evolution of a system without interpretation in terms of any molecular mechanism of chemical reaction. The quantities k_1 and k_{-1} are the *microscopic rate constants* of the mechanism, and they have meaning only in terms of the particular chemical reaction mechanism that has been postulated to account for the rate data.

Let us pause to consider what has been done thus far. We sought a solution for the two differential rate equations, (3.7) and (3.8). Two trial solutions were assumed. Equations (3.15) and (3.16) result, and they are interpreted as constraints placed on the values of the various time-independent constants in order that Equations (3.9) and (3.10) can be valid solutions. Simultaneous solution of Equations (3.15) and (3.16) reveals two roots, named m_0 and m_1, and m_1 is a function of the two k's. The next step is to investigate the implications of the existence of two roots and then to obtain an integrated rate equation that expresses the time dependence of the concentration of reactant or product in this reaction.

Since there are two roots to Equation (3.20), there are two ways to substitute m back into each of the trial solutions (3.9) and (3.10). We shall label the coefficient b or c with a subscript to correspond with the associated m. There result four *particular solutions* of the differential equations:

$$(B) = b_0 e^{-m_0 t} = b_0 e^{-(0)t} = b_0 \tag{3.22}$$

$$(B) = b_1 e^{-m_1 t} \tag{3.23}$$

$$(C) = c_0 e^{-m_0 t} = c_0 e^{-(0)t} = c_0 \tag{3.24}$$

$$(C) = c_1 e^{-m_1 t} \tag{3.25}$$

Neither of the two equations for (B) is valid alone, and the two cannot hold simultaneously. The chemically required result, as well as the mathematically complete result, is the *general solution* obtained by adding together the two particular solutions (3.22) and (3.23) to give

$$(B) = b_0 + b_1 e^{-m_1 t} \tag{3.26}$$

For exactly the same reasons, both Equations (3.24) and (3.25) must be rejected as failing to have chemical meaning if considered individually. The meaningful solution is again the linear combination obtained by adding together solutions (3.24) and (3.25) to give

$$(C) = c_0 + c_1 e^{-m_1 t} \tag{3.27}$$

Equations (3.26) and (3.27) together tell how this chemical system, originally in a nonequilibrium, stressed state, relaxes toward the nonstressed, equilibrium state. Investigation of the relaxation behavior of this chemical system can give a much more detailed and complete picture of the chemistry involved than can consideration of just the final, equilibrium state. Use of these two equations, each with t appearing as the independent variable, permits study of the time dependence of the *approach to equilibrium.*

The time-independent quantities b_0, b_1, c_0, and c_1 have not yet been evaluated. Numerical values of these four quantities depend on the initial concentrations of B and C and on the relative values of the microscopic rate constants. At the beginning of the reaction, at initial time, it is convenient to set t equal to zero. Then[11]

$$(B)_{t=0} = b_0 + b_1 \qquad (3.28)$$

$$(C)_{t=0} = c_0 + c_1 \qquad (3.29)$$

At very large values of t, the reaction approaches a state of chemical equilibrium, and the terms

$$b_1 e^{-m_1 t} \qquad \text{and} \qquad c_1 e^{-m_1 t}$$

individually approach zero. Thus

$$\lim_{t \to \infty} (B) \equiv (\overline{B}) = b_0 \qquad (3.30)$$

$$\lim_{t \to \infty} (C) \equiv (\overline{C}) = c_0 \qquad (3.31)$$

The bar over the concentration denotes the equilibrium value of that concentration. Equations (3.26) and (3.27) can now be written in what will prove to be a useful form:

$$(B)_t - (\overline{B}) = b_1 e^{-m_1 t} \qquad (3.32)$$

$$(C)_t - (\overline{C}) = c_1 e^{-m_1 t} \qquad (3.33)$$

The subscript notation in the quantities $(B)_t$ and $(C)_t$ is a reminder that these concentrations are instantaneous concentrations at time t and that the values of these concentrations change as time changes. Concentrations $(\overline{B})$ and $(\overline{C})$ are time-invariant quantities.

There are three important results of this rate-equation derivation:

1. There exist two microscopic rate constants but only one nonzero macroscopic rate constant. Thus experimental determination of m_1 yields

---

[11] It should be noted that $e^{-(0)} = 1$ and that

$$\lim_{t \to \infty} e^{-mt} = 0$$

the value of neither k_1 nor k_{-1} unless additional information about the relationship between k_1 and k_{-1} is available.

2. The time dependence of each concentration is an exponential function of time. The concentration in each case approaches a limiting value at very large values of t. Thus these kinetic rate equations properly predict the asymptotic approach of the system to a state of chemical equilibrium.

3. The time dependence of each of the concentrations is characterized by the same macroscopic rate constant m_1. The same information about mechanism and about values of individual microscopic rate constants will be obtained if equations such as (3.32) or (3.33) are used to interpret the rate of change of (B) or of (C) or, in fact, of any property of the system which depends linearly on concentration. The time dependence of such a property of the system, and the manner in which the macroscopic rate constant m_1 can be evaluated from a time–property plot, will be explored in the next section.

Experimental Determination of the Macroscopic Rate Constant

It is sometimes possible to follow the time course of the concentration of a single chemical species as the reacting system evolves toward its equilibrium state. Often, though, the most convenient experimental method yields the time dependence of some property of the system which is a function of more than one of the changing concentrations. The experiments described in Chapter 2 did not yield direct information about the time dependence of any particular concentration. It was possible to obtain quantitative data, in terms of the changes in optical absorbance versus time, describing in numerical terms how the color of the solution changed. Such measurements of optical absorbance are especially convenient, because these measurements can be made without disturbing the reacting solution. The problems left unanswered in Chapter 2 were: What do we do with the experimental graphs of absorbance versus time? How can the experimental data be related to a proposed reaction mechanism?

Use of optical absorbance measurements to follow the interconversion of compounds B and C will now be examined. Methods will be developed for obtaining the numerical value of the macroscopic rate constant m_1 from experimental absorbance data obtained at various times during the progress of the reaction. Evaluation of m_1 from this absorbance data is illustrative of general methods that hold for any property of the reacting system which is a linear function [such as Equation (3.34)] of the individual concentrations of reactants and products.

The absorbance, A, of a solution is related to the concentrations of light-absorbing species B and C in solution by the equation

$$A = l\epsilon_B(B) + l\epsilon_C(C) \tag{3.34}$$

where l is the optical path length through the solution and ϵ_B and ϵ_C are constants characteristic of chemical species B and C and are called the molar absorptivities of B and C. The time dependence of the absorbance is obtained by substituting Equations (3.26) and (3.27) into Equation (3.34), giving for the assumed Reaction Mechanism (3.3) the result

$$A_t = l\epsilon_B b_0 + l\epsilon_B b_1 e^{-m_1 t} + l\epsilon_C c_0 + l\epsilon_C c_1 e^{-m_1 t} \qquad (3.35)$$

where A_t is the instantaneous value of A at time t.

Infinite-Time Method.   At times large compared to $1/m_1$, the two exponential terms in Equation (3.35) become insignificantly small in comparison with the two constant terms, and there remains the limiting result

$$\lim_{t \to \infty} A_t \equiv A_\infty = l\epsilon_B b_0 + l\epsilon_C c_0 \qquad (3.36)$$

Combination of Equations (3.35) and (3.36) gives the simplified equation

$$A_t - A_\infty = [l\epsilon_B b_1 + l\epsilon_C c_1] e^{-m_1 t} \qquad (3.37)$$

The specific arrangement of terms was chosen to make use of the facts that

$$\ln[e^x] = x \qquad \text{and} \qquad \ln[a \cdot b] = \ln a + \ln b$$

where ln is an operator that has the meaning "take the logarithm to the base e of the quantity following." A table of logarithms to the base e appears in Appendix III. Taking the logarithm of each side of Equation (3.37) gives

$$\ln |A_t - A_\infty| = \ln |l\epsilon_B b_1 + l\epsilon_C c_1| - m_1 t$$

$$= \text{constant} - m_1 t \qquad (3.38)$$

The absolute-value signs, $|\ |$, appear in Equation (3.38) so that the problem of taking the logarithm of a negative number can be avoided. Inspection of Equation (3.37) reveals that because the exponential must always be positive, both the absorbance difference and the bracketed coefficient of the exponential must be of the same sign. If they are both negative, multiplication of both sides of the equation by -1 makes them both positive. Thus introduction of the absolute-value signs in Equation (3.38) in no way restricts the generality or validity of the subsequent results.

Equation (3.38) has a form that suggests a simple graphical method for determining the value of m_1 from absorbance changes during the course of the chemical reaction. At various times the value of the absorbance is determined, and then the quantity $| A_t - A_\infty |$ is calculated for each value of t. A reliable estimate of the asymptotic limit of the absorbance is required for these calculations. A plot of $\ln | A_t - A_\infty |$ versus t should be a straight line, and the slope of this straight line is the experimental value of $-m_1$. We shall refer to this procedure as the infinite-time method for evaluation of a macroscopic rate constant.

If the experimental points do indeed lie on a straight line, within the limits of experimental error, the assumed mechanism is consistent with experiment. It is then legitimate to associate the measured macroscopic rate constant with the sum of the two microscopic rate constants of the mechanism. If the experimental points do not lie on a straight line, the predictions of the mechanism are not consistent with observation of the actual chemical system, and the mechanism must be rejected. Such a procedure permits mechanisms to be rejected as being inconsistent with experiment, but it does not permit a mechanism to be proved as the only possible mechanism consistent with experiment.

When there is a choice between a simple mechanism and a more complicated mechanism, kineticists have usually taken the position that inclusion of additional intermediates and additional elementary reactions can be justified only when required by experimental evidence. It is always easy to imagine mechanisms of great complexity which are consistent with all data. The criterion which is applied is that, of all mechanisms considered, only the simplest ones consistent with all available experimental data are retained. However, there is no a priori reason to believe that chemical reality is best approximated by the simplest kinetic reaction mechanism. The criterion of simplicity provides a convenient rationale for choosing among alternative mechanisms; it does not necessarily lead toward any sort of ultimate truth.

Guggenheim Method. Use of Equation (3.38) for the graphical evaluation of m_1 is often straightforward, convenient, and accurate. However, in experimental situations in which the asymptotic limit of the property being measured has large uncertainties, this method suffers because of the special emphasis placed on the value of this limit. This problem is avoided in a graphical method, devised by Guggenheim,[12] which we shall apply to the integrated rate equations arising from Reaction Mechanism (3.3).

The quantity $A_{t+\tau}$ is defined by

$$A_{t+\tau} = A \text{ at time } [t + \tau]$$

where τ is an arbitrary but constant interval of time chosen at the con-

[12] E. A. Guggenheim, *Phil. Mag.*, [7] **2**, 538 (1926).

venience of the chemist. Thus we can write Equation (3.35) at time t and also at time $[t + \tau]$ as

$$A_t = l\epsilon_B b_0 + l\epsilon_C c_0 + [l\epsilon_B b_1 + l\epsilon_C c_1]e^{-m_1 t} \tag{3.39}$$

$$A_{t+\tau} = l\epsilon_B b_0 + l\epsilon_C c_0 + [l\epsilon_B b_1 + l\epsilon_C c_1]e^{-m_1[t+\tau]} \tag{3.40}$$

Subtraction of Equation (3.40) from Equation (3.39) produces

$$A_t - A_{t+\tau} = [e^{-m_1 t} - e^{-m_1 t}e^{-m_1 \tau}][l\epsilon_B b_1 + l\epsilon_C c_1]$$

$$= e^{-m_1 t}[1 - e^{-m_1 \tau}][l\epsilon_B b_1 + l\epsilon_C c_1]$$

$$= e^{-m_1 t}[\text{constant}] \tag{3.41}$$

The logarithm to the base e is then taken of each side of Equation (3.41), giving

$$\ln |A_t - A_{t+\tau}| = \ln [\text{constant}] - m_1 t$$

$$= \text{constant} - m_1 t \tag{3.42}$$

Equation (3.42) is the equation of a straight line, and so a plot of

$$\ln |A_t - A_{t+\tau}| \text{ versus } t$$

should yield a straight line with slope equal to $-m_1$. Use of this graphical method does not require a knowledge of the value of A_∞, the absorbance limit at very large values of t.

One way to understand the Guggenheim method is to see it used. Figure 3.1 shows a typical strip-chart recorder trace of absorbance data taken while a reaction was proceeding in solution. The recorder pen moved across the paper, between 0 and 100, in response to changes in absorbance. The paper was being moved through the recorder at a constant rate, in effect converting the edges of the paper (the edges with the sprocket holes) to a time axis. The resulting ink line on the paper becomes a plot of absorbance versus time. Most laboratory recorders have controls that permit the chemist to adjust the range or span of the recorder so that the expected full-absorbance change of the reaction corresponds to about a full chart-paper width, and to adjust the initial pen position to some arbitrary point on the paper such as the extreme left or right side. The recorder trace in Figure 3.1 was obtained by adjusting controls so as to utilize the full width of the chart paper, and there was no attempt made to calibrate the recorder for this kinetic run. We shall see that this procedure, in addition to being very convenient experimentally, introduces no problems in data analysis.

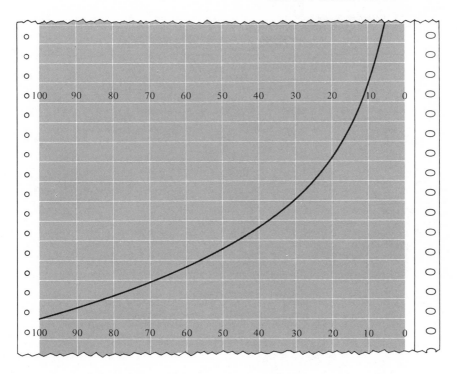

FIGURE 3.1.   Recorder trace of absorbance versus time.

Data must be converted from the ink trace on the chart paper to numerical form suitable for calculations, and this is done by reading ordered pairs of numbers, $[t, A_t]$, from the trace. At selected values of time, the absorbance is read, and these two paired numbers are recorded together in tabular form as shown in Table 3.1. These values of absorbance are in no sense absolute. The observed values differ from absolute absorbance by

$$A_{observed} = \gamma A_{absolute} + \delta \qquad (3.43)$$

where γ and δ have undetermined values which depend on the settings of the recorder controls. Both γ and δ are experimental parameters whose values are adjusted, at the start of the reaction, so as to utilize the full recording range of the strip-chart recorder.

Some convenient value of the constant time interval τ is chosen. In this illustration τ was taken as 160 minutes, or approximately half the time during which the reaction was followed. Simple subtraction gives a value of the quantity $[A_t - A_{t+\tau}]$ at each of the values of t during the first part of the reaction. Note that this subtraction eliminates the undetermined

TABLE 3.1.   Calculations for Graphical Determination of a Macroscopic Rate Constant from Experimental Data: Guggenheim Method

t^*	A_t†	$A_{t+\tau}$‡	$A_t - A_{t+\tau}$	$-\ln|A_t - A_{t+\tau}|$
0§	1.00	0.21¶	0.79	0.24
20	0.82	0.17	0.65	0.43
40	0.67	0.14	0.53	0.63
60	0.55	0.12	0.43	0.84
80	0.45	0.10	0.35	1.05
100	0.37	0.08	0.29	1.24
120	0.31	0.06	0.25	1.39
140	0.25	0.05	0.20	1.61
160	0.21			
180	0.17			
200	0.14			
220	0.12			
240	0.10			
260	0.08			
280	0.06			
300	0.05			

* Minutes.
† Arbitrary units, recorder readings divided by 100.
‡ $\tau = 160$ min.
§ This is not necessarily the time when the reaction began.
¶ This is the value of A_t at $t + \tau$ min $= 160$ min.

constant δ in Equation (3.43). The logarithm to the base e of each absorbance difference is found by use of the tables in Appendix III. Note that the value of the factor γ in Equation (3.43) appears in the resulting column of logarithms as the same constant added to each entry and thus has no effect on the slope of a plot of the logarithms versus time. Entries in the first and last columns of Table 3.1 yield ordered pairs of numbers which are plotted in Figure 3.2. The slope of the straight line drawn through these points is $-m_1$.

The points in Figure 3.2 do not lie exactly on the straight line, and it is really not too surprising to find that there is some scatter of points about the "best" straight line. These points resulted from plotting quantities which were calculated from numbers taken from the recorder chart. There is some uncertainty involved in reading the position of the ink line. In addition, there may be even greater experimental uncertainties arising from the electronic instrumentation or from conditions in the reaction vessel itself. These uncertainties give rise to a corresponding uncertainty in the position of each plotted point in Figure 3.2.

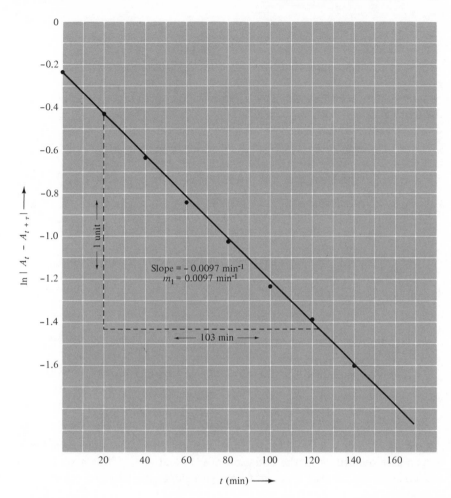

FIGURE 3.2. Graphical determination of a macroscopic rate constant: Guggenheim method.

Time-Lag Method. A convenient graphical method has been described[13] for evaluating the macroscopic rate constant m_1, a method that is closely related to the Guggenheim method but does not require the plotting of logarithms. We shall examine this method for the case of a chemical reaction which proceeds via Mechanism (3.3) and which is being observed by measurement of the change of absorbance of the reacting solution.

[13] E. S. Swinbourne, *J. Chem. Soc.*, **1960**, 2371; P. C. Mangelsdorf, *J. Appl. Phys.*, **30**, 442 (1959).

Again we define the quantity $A_{t+\tau}$ by the equation

$$A_{t+\tau} = A \text{ at time } [t + \tau]$$

where τ is an arbitrary constant time interval. Equation (3.35) can then be written as

$$A_t - [l_{\epsilon B}b_0 + l_{\epsilon C}c_0] = [l_{\epsilon B}b_1 + l_{\epsilon C}c_1] \exp(-m_1 t) \qquad (3.44)$$

$$A_{t+\tau} - [l_{\epsilon B}b_0 + l_{\epsilon C}c_0] = [l_{\epsilon B}b_1 + l_{\epsilon C}c_1] \exp(-m_1[t + \tau]) \qquad (3.45)$$

Division of Equation (3.44) by Equation (3.45) yields

$$\frac{A_t - [l_{\epsilon B}b_0 + l_{\epsilon C}c_0]}{A_{t+\tau} - [l_{\epsilon B}b_0 + l_{\epsilon C}c_0]} = \exp(-m_1[t - t - \tau]) = \exp(m_1\tau)$$

Algebraic rearrangement gives

$$A_t - [l_{\epsilon B}b_0 + l_{\epsilon C}c_0] = A_{t+\tau} \exp(m_1\tau) - [l_{\epsilon B}b_0 + l_{\epsilon C}c_0] \exp(m_1\tau)$$

The time-independent terms are now collected, giving the result

$$A_t = \{l_{\epsilon B}b_0 + l_{\epsilon C}c_0\}\{1 - e^{m_1\tau}\} + A_{t+\tau} \exp(m_1\tau) \qquad (3.46)$$

Equation (3.46) is the equation of a straight line. The only two time-dependent quantities in the equation are A_t and $A_{t+\tau}$. Thus a plot of A_t versus $A_{t+\tau}$ should give a straight line with slope equal to

$$e^{m_1\tau}$$

The logarithm to the base e of the slope is equal to $m_1\tau$. The numerical value of the constant-time interval τ is known, so the value of the macroscopic rate constant m_1 can be calculated directly.

The plotting of the experimental data is especially straightforward. The only place where a table of logarithms is needed is at the very end, when the slope is converted to the sought-for macroscopic rate constant. Another desirable feature of this graphical method is the fact that the slope of the line is influenced most by the most reliable data, least by the least reliable data. Data from the earliest part of the reaction are spread out along the line, and these data dominate the plot. The numbers in columns

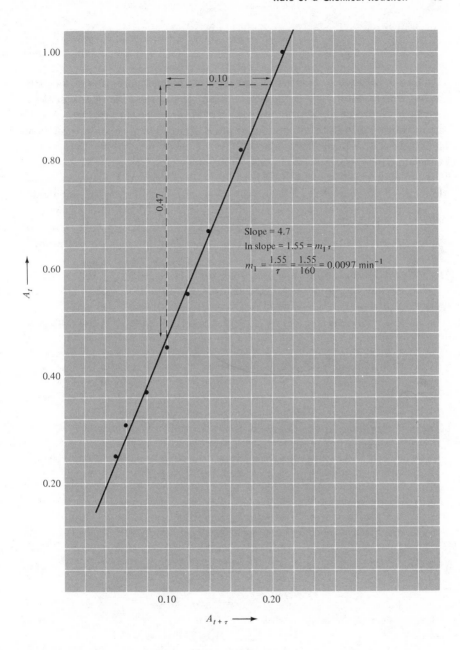

Slope = 4.7

ln slope = 1.55 = $m_1 \tau$

$m_1 = \dfrac{1.55}{\tau} = \dfrac{1.55}{160} = 0.0097 \text{ min}^{-1}$

FIGURE 3.3. Graphical determination of a macroscopic rate constant: time-lag method.

two and three in Table 3.1 can be used directly to construct a time-lag plot. Such a plot is presented in Figure 3.3.

Half-Life Method.   There is a fascinating property of the exponential curve which results when Equation (3.32) is plotted. There is a sense in which it is correct to say that the shape of this curve stays the same all the way from zero time to unboundedly great times. Let us choose some arbitrary time t and note the value of (B) at this time, calling that instantaneous concentration value $(B)_t$. How long will it take for the value of (B) to change halfway from $(B)_t$ to $(\overline{B})$? This interval of time, called the half-life of the reaction and denoted by $t_{1/2}$, turns out to have the same value no matter what point on the curve is chosen for t. The half-life is characteristic of the reaction, and its numerical value is closely related to the value of m_1. This fact permits experimental evaluation of m_1 by determination of the half-life.

By definition, the time required for $[(B)_t - (\overline{B})]$ to change to $\frac{1}{2}[(B)_t - (\overline{B})]$ is equal to $t_{1/2}$. Let us evaluate Equation (3.32) at $t = t_{1/2}$:

$$\tfrac{1}{2}[(B)_t - (\overline{B})] = \tfrac{1}{2}b_1 e^{-m_1 t} = b_1 e^{-m_1[t+t_{1/2}]} \tag{3.47}$$

From Equation (3.47) we can see that

$$\tfrac{1}{2}e^{-m_1 t} = e^{-m_1[t+t_{1/2}]} \tag{3.48}$$

Taking advantage of the fact that

$$0.500 = e^{-0.693} \qquad \text{and} \qquad e^{\alpha} \cdot e^{\beta} = e^{[\alpha+\beta]}$$

we write Equation (3.48) as

$$e^{-0.693} \cdot e^{-m_1 t} = e^{-[0.693+m_1 t]} = e^{-m_1[t+t_{1/2}]} \tag{3.49}$$

The two bracketed exponents in Equations (3.49) must be equal, and so

$$0.693 + m_1 t = m_1 t + m_1 t_{1/2}$$

or

$$t_{1/2} = \frac{0.693}{m_1} \tag{3.50}$$

The same derivation could have been performed with Equation (3.33) as a starting point. [Equation (3.37) would have served equally well.]

The half-life is a characteristic of the reacting system. Its value can be determined simply by finding how long it takes for the value of any concentration, or the value of any property that is a linear function of the concentrations, to change half-way to its equilibrium value.

It is easy to evaluate the half-life from a time–property graph if a reliable equilibrium value of the property can be obtained. But there is a pitfall. The concept of a unique half-life for a reaction only has validity if the time–property graph is an exponential function of the form of (3.32), (3.33), or (3.37). There needs to be some verification that the experimental graph indeed has the proper form. One check is to determine the half-life

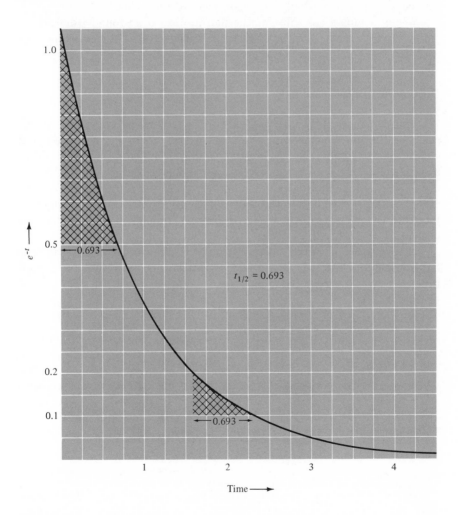

FIGURE 3.4. Determination of half-lives from an exponential curve.

at several different places along the curve; if the curve is an exponential curve, the value of the half-life will be independent of the region of the curve used for its evaluation. Verification of the functional form of the rate curve is important, because many other time–property curves are frequently encountered. In Chapter 5 we find that a mechanism of coupled unimolecular reactions gives rise to a time–property curve that is written as a sum of exponential terms. Chapters 7, 8, 9, and 10 deal with mechanisms that contain at least one bimolecular elementary process; there are experimental conditions under which such reactions yield time–property curves that are sums of exponentials, and other conditions under which the rate curves are substantially more complicated functions.

When a numerical value of the half-life has been found from an experiment, Equation (3.50) is used for calculation of the macroscopic rate constant m_1.

Differential Rate Equations

Another method of analyzing a chemically reacting system for evaluation of the microscopic rate constants makes direct use of differential equations (3.7) and (3.8). Tying the discussion of the general method to a specific case, we shall again examine a reaction that proceeds via Mechanism (3.3) and is followed experimentally with a recording spectrophotometer. Instead of looking at the value of A at various times, we shall be interested in the *slope* of the curve of A versus time. This slope is the time derivative of A, dA/dt.

The absorbance of the reacting solution is given at each instant in time by Equation (3.34), where the concentrations (B) and (C) are instantaneous concentrations. Differentiation[14] of Equation (3.34) with respect to time yields

$$\frac{dA}{dt} = l\epsilon_B \frac{d(B)}{dt} + l\epsilon_C \frac{d(C)}{dt} \tag{3.51}$$

where each of the derivatives is an instantaneous derivative, a derivative

---

[14] The time derivative of the product, ax, of a time-independent constant a and a time-dependent variable x is given by the expression

$$\frac{d[ax]}{dt} = a \frac{dx}{dt}$$

The derivative of the sum of several terms is the sum of the derivatives of the individual terms.

whose value changes continuously with time throughout the course of the reaction. For the specific mechanism under consideration, the derivatives on the right-hand side of Equation (3.51) are given by differential equations (3.7) and (3.8). Substitution into Equation (3.51) gives

$$\frac{dA}{dt} = k_1 l [\epsilon_C - \epsilon_B](B) + k_{-1} l [\epsilon_B - \epsilon_C](C)$$

$$= l [\epsilon_C - \epsilon_B][k_1(B) - k_{-1}(C)] \tag{3.52}$$

Equation (3.52) is not very useful for determining the values of the k's from absorbance measurements taken throughout the reaction, because the observed absorbance is written as a function of two instantaneous concentrations, neither one of which may be readily measured as the reaction progresses. However, if it is possible to start a kinetic experiment with just B present and thus with an initially zero concentration of C, then a limiting form of Equation (3.52) can be used to evaluate the microscopic rate constant k_1. Such a kinetic run is called an initial-rate experiment.

Initial-Rate Method. If a solution initially contains B but not C, then at the beginning of the reaction, $(C) = 0$, and Equation (3.52) becomes

$$\left(\frac{dA}{dt}\right)_0 = l [\epsilon_C - \epsilon_B] k_1(B)_0 \tag{3.53}$$

where $(B)_0$ is the initial concentration of B and $(dA/dt)_0$ is the initial slope of a plot of A versus time. If the numerical value of each of the quantities l, ϵ_B, ϵ_C, and $(B)_0$ is known independently of the kinetic measurements, k_1 can be evaluated from measurement of the initial rate of the reaction.

If a solution initially contains C but not B, then at the beginning of the reaction $(B) = 0$, and Equation (3.52) becomes

$$\left(\frac{dA}{dt}\right)_0 = l [\epsilon_B - \epsilon_C] k_{-1}(C)_0 \tag{3.54}$$

where $(C)_0$ is the initial concentration of C and $(dA/dt)_0$ is the initial slope of a plot of A versus time. A knowledge from independent measurements of the numerical values of l, ϵ_B, ϵ_C, and $(C)_0$ is required if k_{-1} is to be determined from such an experiment.

Use of the initial-rate method requires extrapolation of the absorbance–time curve back to the time when chemical reaction was initiated. The slope of the curve at this initial time is determined by drawing a straight line, tangent to the curve at the point where $t = 0$. The slope of this tangent line is the initial slope of the curve. The ideal situation occurs when the initial portion of the absorbance–time curve is essentially linear. Then the slope of this linear portion is the initial rate of the reaction. Use of Equation (3.53) requires an initial-rate region in which measurable absorbance changes occur without significant change in the value of the concentration (B). The inequality $k_1(B) \gg k_{-1}(C)$ must hold throughout the initial-rate region. There is not always such a region which is experimentally accessible, and the initial-rate method is therefore not always applicable.

Comparisons of initial-rate, infinite-time, half-life, Guggenheim, and time-lag methods reveal substantial differences:

The initial-rate method is based on differential rate equations. Infinite-time, Guggenheim, time-lag, and half-life methods are based on integrated equations derived from the differential rate equations.

The initial-rate method can in favorable cases give individual numerical values for both k_1 and k_{-1}. The methods based on integrated equations yield only the sum of microscopic rate constants.

Initial-rate calculations require a rather detailed knowledge of chemical constants such as initial concentrations and individual species absorptivities. The infinite-time, Guggenheim, time-lag, and half-life calculations require quite minimal characterization of the time-independent quantities.

Initial-rate experiments must start with a reaction mixture of carefully specified concentrations. The mixture must begin to react at some known time. The other methods thus far discussed, based on integrated rate equations, have no such restrictions.

Methods based on integrated equations utilize data from throughout the entire time of the reaction. (This is true of the half-life method if $t_{1/2}$ is evaluated in several different regions of the rate curve.) In the initial-rate method, the numerical value obtained for the rate constant depends on data from just the earliest part of the reaction.

The Guggenheim, time-lag, and infinite-time methods provide checks on the predictions of the mechanism, since the mechanism requires that certain plots be linear. A curved plot implies an invalid mechanism or poor experiments. The half-life method requires that the value of $t_{1/2}$ be independent of the region of the rate curve used in its evaluation. There is no corresponding check for use with the initial-rate method.

Rate Constants and the Equilibrium Constant

The apparent[15] equilibrium constant, K', for the reaction

$$B \rightleftharpoons C$$

is written in terms of equilibrium molar concentrations as

$$K' = \frac{(\overline{C})}{(\overline{B})} \tag{3.55}$$

Equation (3.55) is meaningful only for a system in a state of chemical equilibrium, a system in which no concentration is changing with time. A requirement imposed by the existence of chemical equilibrium is

$$\frac{d(B)}{dt} = 0 \tag{3.56}$$

$$\frac{d(C)}{dt} = 0 \tag{3.57}$$

If the kinetic reaction mechanism by which chemical equilibrium is attained is (3.3), then Equations (3.7) and (3.8) must be valid at all values of t. In particular, they must hold at equilibrium. Substitution of Equation (3.56) into Equation (3.8), or substitution of Equation (3.57) into Equation (3.7), yields

$$k_1(\overline{B}) = k_{-1}(\overline{C}) \tag{3.58}$$

which can be rearranged into the form suggested by the equilibrium constant equation,

$$\frac{k_1}{k_{-1}} = \frac{(\overline{C})}{(\overline{B})} = K' \tag{3.59}$$

A specific kinetic reaction mechanism has been assumed, and then an

---

[15] The thermodynamic equilibrium constant, K, is properly written as the quotient of *activities* of reactant and product. It is possible, and it is general practice, to define the activities in such a way so that, for very dilute solutions, the value of the activity of each solute species approaches the value of the concentration of that species. Another quantity, the apparent equilibrium constant, K', is defined as the quotient of molar concentrations of reactant and product. In the limit of infinite dilution, $K = K'$. For dilute solutions, the two equilibrium constants usually have nearly the same value. Substantial divergence between the two numerical values is to be expected for concentrated solutions. For any particular solution, the quantity K' has a definite and time-independent value at constant T and P.

equation was derived to relate the two microscopic rate constants of the mechanism to the one equilibrium constant for the reaction. If a different kinetic mechanism had been assumed, then some different relationship between rate constants and the equilibrium constant would have resulted. The numerical value of the equilibrium constant is independent of the mechanism by which equilibrium is achieved, but if the mechanism were known and each of the rate constants evaluated, the value of the equilibrium constant could be calculated.

It is a general result that information obtained from the measurement of equilibrium properties of a chemical system can give no information about the chemical reaction pathway by which some nonequilibrium system in the past evolved to become the present equilibrium system. There is no way for an equilibrium system to contain a memory of times past.

Problems

3.1. To get from Equation (3.19) to Equation (3.20), it is necessary to divide by c. It is therefore essential that the value of c be different from zero. Show that c is, in fact, always nonzero for all cases of interest for reaction-rate studies.

3.2. Verify by substitution that Equations (3.26) and (3.27) are solutions of the two simultaneous differential rate equations (3.7) and (3.8).

3.3. If a reaction is proceeding according to the mechanism given in (3.3), the *relaxation time* of the reaction is the time required for the difference between the instantaneous value of any concentration and its equilibrium value to decrease by the factor $1/e$. Calling the relaxation time τ, find a relationship between τ and m_1.

3.4. Some people prefer to use base-10 logarithms for plotting experimental data when using the Guggenheim or the infinite-time methods. Verify that a straight-line plot will result, and find the relationship between the slope of such a plot and the numerical value of m_1.

3.5. Show that the time, $t_{1/n}$, required for a reaction following Mechanism (3.3) to get from its initial state to a state $1/n$th of the way to equilibrium is given by

$$t_{1/n} = \frac{1}{m_1} \ln \left(\frac{1}{1 - \frac{1}{n}} \right)$$

3.6. The reaction in water of pyridoxal (a compound closely related to the coenzyme vitamin B_6) and the amino acid alanine is accompanied by the appearance of yellow color in the solution. The reaction can be initiated by mixing a solution of pyridoxal with a solution of alanine. (The concentration of alanine, 50 times greater than the concentration of pyridoxal in this illustrative kinetic run, can be considered constant. This is an example of the constant-concentration approximation, which is discussed in Chapter 8.) Use the following student data to evaluate, by the Guggenheim method, the macroscopic first-order rate constant m_1:

t (sec)	Absorbance	t (sec)	Absorbance
30	0.722	420	1.583
60	0.850	450	1.603
90	0.967	480	1.621
120	1.071	510	1.633
150	1.164	540	1.645
180	1.242	570	1.656
210	1.309	600	1.667
240	1.367	630	1.675
270	1.421	660	1.682
300	1.464	690	1.688
330	1.500	720	1.694
360	1.529	750	1.699
390	1.559		

3.7. Conventionally, a chemist sets t equal to zero at the beginning of a kinetic run. This is an arbitrary action. Show that Equations (3.32) and (3.33) have the same form regardless of whether $t = 0$ or $t = \tau$ at the beginning of the reaction.

BIBLIOGRAPHICAL NOTE

The mathematical model associated with Mechanism (3.3) can be formulated in the language of probability theory, giving rise to equations equivalent to (3.26) and (3.27). This probabilistic model of reaction rates is discussed in detail in D. A. McQuarrie, *Stochastic Approach to Chemical Kinetics*, London: Methuen & Co., Ltd., 1967. Probability arguments can be used in another way to set up a mathematical model for a reacting chemical system. The Monte Carlo method, involving consideration of the statistics of random events, can be employed for plotting the time course of a chemical reaction. An introduction to the use of Monte Carlo techniques in chemical kinetics is given by B. Rabinovitch, *J. Chem. Educ.*, **46,** 262 (1969).

The exponential function appears frequently in chemistry. We encountered it in this chapter as part of trial functions (3.9) and (3.10).

Had we wished to derive the Bouguer–Beer law in Chapter 2 by using calculus, we would have encountered the exponential function there. Some of the chemical applications of this mathematical function are explored in articles by S. Y. Shen, S.-R. Huang, and S. Karp [*Chemistry*, **42**(2), 16 (1969)] and by F. O. Green [*J. Chem. Educ.*, **46**, 451 (1969)]. See also a note by C. M. Crawford [*Chemistry*, **42**(6), 32 (1969)].

Most of this chapter is devoted to the use of a mathematical model of a chemical reaction, to finding ways that make possible an informative confrontation between the results of an experiment and the predictions of a mechanistic model. An excellent motion picture, illustrating the use and meaning of mathematical models in understanding the physical world, is *The Mathematician and the River* (16 mm, color, 19 minutes, sound). Flood control on the Mississippi River is used as the example. The film was produced by Educational Testing Service, Princeton, N.J., as part of their Horizons of Science film series; E.T.S. will supply a list of loan centers.

4

Racemization of an Asymmetric Complex Ion:

A Case Study

The *cis*-dichloro-bis(ethylenediamine)chromium(III) complex ion, which we will call *cis*-[Cr en$_2$Cl$_2$]$^+$, exists in two forms, which differ only in the arrangement of the atoms in space. The geometry of these two forms differs in much the same way as the geometry of the left hand differs from the geometry of the right hand of the same person. Just as the left hand is the mirror image of the right hand, so the two isomers of *cis*-[Cr en$_2$Cl$_2$]$^+$ cation are mirror images of each other. The two isomers, differentiated in the following discussion by the prefixes *l*- and *d*-, have the three-dimensional structures

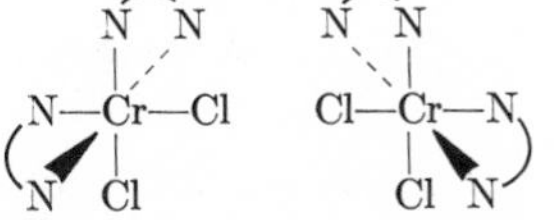

$$
\left(
\begin{array}{c}
\overset{\frown}{\text{N}\ \text{N}} \\
| \\
\text{N—Cr—Cl} \\
\nearrow\ | \\
\text{N}\ \ \text{Cl}
\end{array}
\qquad
\begin{array}{c}
\overset{\frown}{\text{N}\ \text{N}} \\
| \\
\text{Cl—Cr—N} \\
|\ \searrow \\
\text{Cl}\ \ \text{N}
\end{array}
\right)
\tag{4.1}
$$

where N⌒N indicates the ligand ethylenediamine, H$_2$N—CH$_2$—CH$_2$—NH$_2$.

When the compound *l*-*cis*-[Cr en$_2$Cl$_2$]Cl·H$_2$O is dissolved in water, a variety of chemical changes occur. These changes take place with rates such that observations made during times of a few minutes or a few hours can yield a great deal of chemically meaningful experimental information. The concentration of the anion Cl$^-$ increases with time. The color of the solution changes with time. There is also a change in a quantity called the *optical rotation* of the solution. Not all of these changes occur at the same rate. The optical rotation changes can, in fact, be resolved into two distinct changes. Enough independent information seems to be accessible from rate experiments to justify some detailed conclusions about just what is happening in solution and about the reaction mechanism by which the chemical transformations are brought about.

Before discussing specific details of this reaction, some aspects of the measurement of optical rotation will be examined.

Measurement of Optical Rotation.   Associated with any beam of light is an electromagnetic disturbance along the path of the light beam. An electric field vibrates in a direction that is perpendicular to the path of the beam. There is an infinity of such directions. The electric field accompanying the light that radiates from the tungsten filament of an ordinary electric light bulb oscillates in a random and complicated way, without pattern or order. However, when such an ordinary light beam passes through a crystal of a doubly refracting material (calcite is such a material), a surprising phenomenon occurs. Inside that crystal, the original beam becomes two beams. The two beams have two different speeds and two different directions, and at each instant the electric-field vector[1] associated with one of the beams is perpendicular to the corresponding vector associated with the other beam. One beam is called the O ray (the ordinary ray), and the other is called the E ray (the extraordinary ray). In each beam, the electric-field vector vibrates in a particularly simple, orderly manner. Because of the simple pattern of the vibration, each beam is said to be *polarized*. Figure 4.1 shows schematically why the beam is described as being *linearly* polarized. The vibration of the electric vector is linear. This vibrating electric vector sweeps out a plane as the polarized beam passes through empty space or through some uniform medium.

There is an alternative way of representing the vibration of this electric vector. A way that is useful in talking about optical rotation is illustrated in Figure 4.2. Here the linearly vibrating vector is considered as the sum of two rotating vectors. A special situation will now be investigated in which interactions of a clockwise-rotating electric vector with the molecules in the path of the beam are different from such interactions of a counterclockwise-rotating vector.

There is a fundamental relationship between the clockwise and counterclockwise natures of these rotating vectors, and the right-handed and left-handed natures of such compounds as the two isomers (4.1). If a beam

---

[1] A vector is drawn on paper as an arrow, and it has both a length and a direction. The electric vector represents the electric field associated with the light beam. The length of this vector at a particular point is proportional to the instantaneous magnitude of the electric field at that point. The vector points in the direction of the field. The electric vector associated with a light beam is constantly changing, thus representing the rapidly oscillating electric field.

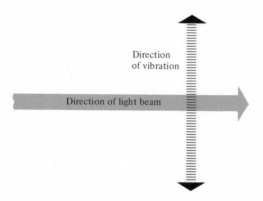

Direction
of vibration

Direction of light beam

FIGURE 4.1. Representation of a beam of linearly polarized light. The electric vector vibrates in a direction perpendicular to the direction of propagation of the light beam. If the beam is passing through empty space, these two lines sweep out a plane as the light passes through the medium. If the beam is passing through a solution which contains unequal populations of asymmetric molecules, the electric vector will be rotated as the light beam moves through the solution. The angle by which the electric vector is rotated around the light beam axis is proportional to the distance the light beam has traveled through the solution and to the number of asymmetric molecules in the path of the light beam.

of linearly polarized light is passed through a solution containing unequal numbers of the two isomeric molecules, the velocity of the clockwise vector will be different from the velocity of the counterclockwise vector. One vector will thus progress faster along the direction of the beam than will the other vector. The sum of the two rotating vectors at any point along the beam will still be a linearly vibrating vector, but the direction of this resultant vector will have been rotated about the beam. The solution containing asymmetric molecules will have rotated the plane of polarization of the light by α angular degrees. Such a solution is said to be *optically active*.

Quantitative measurement of optical rotation of a solution is made with a polarimeter. The essential optical components of a polarimeter are shown in Figure 4.3. A common form of polarizer is a prism cut from doubly refracting calcite. This prism is made in such a way that the extraordinary rays are refracted out of the optical path of the instrument. The analyzer can be a similar prism, mounted so that it can be rotated around the light-beam axis. When the polarizer and analyzer prisms are aligned perpendicular to one another (analyzer rotated 90° around the light-beam axis), the light transmitted through the analyzer is at a minimum. A slight rotation of the analyzer in either direction increases the intensity of light pass-

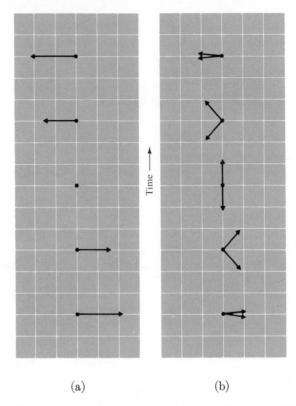

FIGURE 4.2. End-on views of a beam of linearly polarized light. (a) Electric vector considered as always in the plane of polarization, with the length varying periodically with time. (b) Electric vector considered as sum of two rotating vectors, one vector rotating clockwise, the other rotating counterclockwise, at the same rate. Length of each vector remains constant.

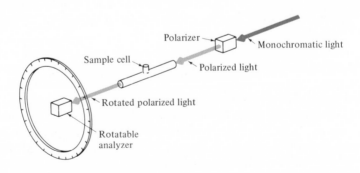

FIGURE 4.3. Components of a polarimeter.

ing through it. The angular position which yields minimum light intensity is called the null point.

The analyzer is attached to a rigid metal circle that is accurately divided in degrees and fractions of degrees so that the angular position of the analyzer can be determined. Precision polarimeters are available which permit this angular position to be read with a precision of $\pm 0.001°$. When making measurements, a null-point reading is taken with the sample cell filled with solvent, and again with the sample cell filled with solution containing optically active molecules. The difference between the two readings is the optical rotation, α.

Insofar as each asymmetric molecule can be considered to rotate polarized light independently of all other molecules in solution, the total rotation will be proportional to the number of molecules in the path of the beam of polarized light. For a solution containing the optically active compound, X, the optical rotation is given by

$$\alpha = l\alpha_X(X) \tag{4.2}$$

where α is the observed rotation; (X) is the concentration of the only asymmetric species, X; α_X is the proportionality constant, called the molar rotatory coefficient of species X; and l is the optical path length through the solution. The molar rotatory coefficient is a function of wavelength of the light used for the measurement.

If there are two different optically active chemical species in solution, then Equation (4.2) becomes

$$\alpha = l\alpha_{X_1}(X_1) + l\alpha_{X_2}(X_2) \tag{4.3}$$

In general, for a solution containing n different and independent optically active species, the observed rotation is given by

$$\alpha = l\sum_{i=1}^{n} \alpha_{X_i}(X_i) \tag{4.4}$$

where the $\sum$ indicates summation of a series of n terms. The first term of the series has $i = 1$ and is $\alpha_{X_1}(X_1)$. The second term, with $i = 2$, is $\alpha_{X_2}(X_2)$. The final term, with $i = n$, is $\alpha_{X_n}(X_n)$.

For the particular case of two mirror-image isomers, the values of the molar rotatory coefficients must be equal in magnitude but opposite in sign:

$$\alpha_{l-X} = -\alpha_{d-X} \tag{4.5}$$

Equation (4.5) requires that an equimolar mixture of mirror-image isomers have a rotation of zero degrees. Such a mixture is called a racemic mixture. The racemic mixture is the equilibrium distribution of the two chemically identical but geometrically different molecules. The chemical reaction that results in interconversion of the two mirror-image isomers is called a racemization.

Measurement of optical rotation is fundamentally a simple operation if a good polarimeter is available. There are a few hidden sources of error which should be known, including variations in temperature within the cell and the presence of dirt in the solution and on the ends of the cell. It is wise to use a thermostatted sample cell, not only because of the effect of temperature on reaction rates if a reaction-rate curve is being determined, but also because temperature variations along the length of the cell can make accurate measurements almost impossible. Cleanliness is very important. Dirt particles suspended in the solution may rotate the polarized light, as may dust, dirt, and fingerprints on the optical surfaces of the cell.

Preliminary Spectrophotometric Experiment. A solution was prepared by dissolving 0.015 gram of *cis*-[Cr en$_2$Cl$_2$]Cl in distilled water. The optically active salt was synthesized by the procedure described at the end of this chapter. The solution was made up to 50 ml in a volumetric flask. These crystals dissolve readily to produce a pink solution. A portion of the solution was transferred to a cylindrical spectrophotometer cell having a a 100-mm optical path length, and the absorption spectrum was recorded using a Perkin-Elmer Model 350 spectrophotometer. The spectrum was obtained within 5 minutes after the solid was added to water, and then again after total elapsed times of 20, 45, and 80 minutes. These spectra are presented together in Figure 4.4.

Significant changes in the absorption spectrum occur within this time period. In an experiment designed to determine a reaction-rate curve, a thermostatted solution would be allowed to react in the cell compartment of the spectrophotometer. The absorbance would be recorded continuously at an appropriate wavelength. Selection of the appropriate wavelength can be made by inspection of the spectra in Figure 4.4. A wavelength can be picked at which absorbance will increase with time. A different wavelength can be found at which the absorbance will decrease with time. There are also wavelengths at which no change will be observed.

Preliminary Polarimetric Experiment. A solution was prepared by dissolving 0.014 gram of *l-cis*-[Cr en$_2$Cl$_2$]Cl in distilled water. The solution was made up to 25 ml in a volumetric flask. A portion of the solution was transferred to a polarimeter cell having a 100-mm optical path length. This polarimeter cell is surrounded by a jacket through which thermostatted

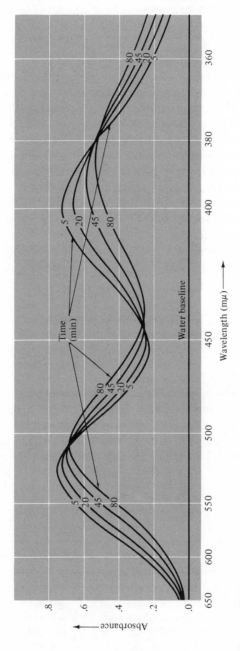

FIGURE 4.4.   Spectra of the reacting solution.

water from a constant-temperature bath was circulated. The temperature was maintained at 25.0°C. Measurements of optical rotation were made 10 minutes after the solid was added to the water, and again 30 minutes after the solid was added, using a Perkin-Elmer Model 141 polarimeter. Data from these measurements are presented in Table 4.1.

Even these limited data are sufficient to show that the reaction is not simply a racemization, because optical rotation during a simple racemization must continuously approach zero. For this reaction, the rotation observed using light of wavelength 436, 365, and 313 mμ in each case changed away from zero during the first 30 minutes. If racemization is occurring simultaneously with some other reaction, the value of the optical rotation at these three wavelengths must pass through a maximum at some time, finally tending toward zero at large values of t. The observed reaction-rate curve could then be considered as the superposition of two relaxation processes. Some mechanisms that give rise to two superposed relaxations are discussed in Chapter 5.

These same data also show that the reaction is not simply a conversion of the *cis* form of the chromium(III) complex cation to the *trans* form:

$$
\begin{array}{ccc}
& \overset{N \frown N}{\underset{N}{\overset{|}{N-Cr-Cl}}} \rightleftharpoons \overset{Cl\ N}{\underset{N}{\overset{|}{N-Cr-N}}} \\
& cis \qquad\qquad trans &
\end{array}
\tag{4.6}
$$

The mirror image of the *trans* isomer is identical with the original molecule. Left- and right-handed forms of this molecule are not distinguishable, so a solution of the *trans* isomer will not rotate polarized light. Thus if Reaction (4.6) were the sole step in the reaction mechanism, the optical rotation would be required to approach zero throughout the reaction.

TABLE 4.1.   Optical Rotation Measurements

Wavelength (mμ)	Optical Rotation (deg)	
	After 10 min	After 30 min
589	+0.113	+0.079
546	+0.338	+0.282
436	−0.073	−0.116
365	−0.118	−0.135
313	−0.157	−0.173

Speculation Concerning Mechanism. It has been suggested[2] that there are least four reactions occurring simultaneously in this solution, including the two aquation reactions

$$[Cr\ en_2Cl_2]^+ + H_2O = [Cr\ en_2ClH_2O]^{2+} + Cl^-$$

$$[Cr\ en_2ClH_2O]^{2+} + H_2O = [Cr\ en_2(H_2O)_2]^{3+} + Cl^-$$

and the two racemization reactions

$$l\text{-}[Cr\ en_2ClH_2O]^{2+} = d\text{-}[Cr\ en_2ClH_2O]^{2+}$$

$$l\text{-}[Cr\ en_2Cl_2]^+ = d\text{-}[Cr\ en_2Cl_2]^+$$

There is also the possibility of *cis-trans* isomerization. Several mechanistic possibilities exist for each of these reactions. Analogies are drawn by Selbin and Bailar with the corresponding cobalt(III) complexes, which have been studied and interpreted in detail.[3,4]

SUGGESTION FOR EXPERIMENTAL INVESTIGATION

Design an experiment for the determination of two macroscopic rate constants from polarimetric data. Compare the results with data obtained from spectrophotometric rate experiments. What is the simplest kinetic reaction mechanism consistent with all your experimental observations? Are there any other reasonable mechanisms that should be considered? Propose an experimental method that might distinguish between two plausible mechanisms. If time and equipment are available, carry out the required experiments.

Preparation of Optically Active l-cis-Dichloro-bis (ethylenediamine)Chromium(III) Chloride

Experimental investigation of the rate processes described in this case study can be coupled in an undergraduate laboratory with the synthesis and resolution of the optically active reactant. Portions of the preparation can be conducted by a pair of students, or responsibility for each of the subpreparations can be delegated to groups of students, in order to use

[2] J. Selbin and J. C. Bailar, Jr., *J. Amer. Chem. Soc.*, **79,** 4285 (1957).

[3] F. Basolo and R. G. Pearson, *Mechanisms of Inorganic Reactions: A Study of Metal Complexes in Solution*, New York: John Wiley & Sons, Inc., 2nd ed., 1967, chap. 4.

[4] C. H. Langford and H. B. Gray, *Ligand Substitution Processes*, New York: W. A. Benjamin, Inc., 1965, chap. 3.

a minimum of specialized equipment. The preparation has been adapted from procedures described by Selbin and Bailar[5] and Rollinson and Bailar.[6]

Preparation of [Cr en₃]₂(SO₄)₃. The salt chromium(III) sulfate and the liquid ethylenediamine are allowed to react, the reaction being described by the equation

$$Cr_2(SO_4)_3 + 6en = [Cr\ en_3]_2(SO_4)_3$$

Heat together stoichiometric amounts of anhydrous chromium(III) sulfate (24.5 grams) and anhydrous ethylenediamine (25 ml) on a steam cone in a 500-ml wide-mouth round-bottom flask with a ground-glass joint. Insert an air condenser with a ground-glass joint in the mouth of the flask so that volatilized ethylenediamine is condensed and returned to the re-action flask. The flask must be clamped in place on the steam cone so that the assembly does not topple over. Ethylenediamine is very corrosive and must be kept off skin and clothing.

As the chemical reaction proceeds, the suspension loses its bright green color. If this does not happen within 1 hour, add a drop of water to increase the reaction rate. Shake the reaction assembly occasionally to keep the solids suspended and thus exposed to unreacted ethylenediamine. A brown mass is eventually formed, and no liquid remains. Leave the flask heating on the steam cone overnight to complete the reaction. Care must be taken to exclude water from the reaction mixture. If more than a trace of water is present, the reaction may not go to completion. One troublesome source of water can be condensed steam from the steam cone. An inverted beaker covering the top of the air condenser, or a $CaCl_2$ drying tube inserted in the end of the air condenser, have proved helpful.

Remove the mass of product from the flask, using a Teflon policeman to avoid scratching the glass flask. Grind this product to a powder with a mortar and pestle. Suspend the powder in ethanol and then collect this solid by vacuum filtration through filter paper in a Büchner funnel. Finally, spread the product on a watch glass to dry.

Preparation of [Cr en₃]Cl₃·3.5H₂O. The sulfate salt is converted to the chloride by dissolving [Cr en₃]₂(SO₄)₃ in hydrochloric acid and then pre-cipitating the chloride in the presence of an overwhelming excess of chloride ion by the addition of ethanol.

[5] See footnote 2.

[6] C. L. Rollinson and J. C. Bailar, Jr., *Inorganic Syntheses*, vol. II, New York: McGraw-Hill, Inc., 1946, pp. 196–202.

Prepare some hydrochloric acid solution by mixing 6 volumes of distilled water and 1 volume of concentrated hydrochloric acid. For each gram of $[Cr\ en_3]_2(SO_4)_3$, take 1 ml of the hydrochloric acid solution and heat to 60 to 65°C. Dissolve the sulfate salt in the hot hydrochloric acid solution and then immediately filter through filter paper in a Büchner funnel. *Save the filtrate!* For each milliliter of filtrate, add 2 ml of a solution made up from 9 volumes of concentrated hydrochloric acid and 14 volumes of ethanol. Cool the resulting solution, with stirring, in ice. Collect the yellow crystals of the chloride salt in the Büchner funnel.

This product is contaminated with sulfate. Thermal decomposition in the next step is catalyzed by traces of chloride, and it is desirable to have the crystals contaminated with a volatile chloride. Removal of sulfate ions and introduction of some ammonium chloride into the product are both accomplished by recrystallization from a dilute ammonium chloride solution. Make up a 5 percent (weight percent) solution of ammonium chloride in water. Take 1 ml of solution for each gram of product, heat to 65°C, dissolve product, and cool. Collect these crystals in the Büchner funnel. Spread the crystals on a large watch glass to dry.

Preparation of *cis*-[Cr en₂Cl₂]Cl·H₂O. Thermal decomposition of $[Cr\ en_3]Cl_3$ proceeds, under the conditions used for this synthesis, largely according to the reaction

$$[Cr\ en_3]Cl_3 \triangleq cis\text{-}[Cr\ en_2Cl_2]Cl + en \uparrow$$

A molecule of ethylenediamine leaves the inner coordination sphere and is replaced by two chloride ions which had been nearby in the outer coordination sphere.

Place dried crystals of $[Cr\ en_3]Cl_3 \cdot 3.5H_2O$, spread in a thin layer on the large watch glass, in an oven at 210°C. Careful control of temperature is essential. Other decomposition processes begin to compete with the desired reaction at temperatures above 215°C. Below 200°C the reaction is very slow. Evolution of ethylenediamine gas begins within a few minutes, and after 1 to 2 hours the salt has become red-violet. About 85 percent conversion can be expected during this period of time.

Dissolve the red-violet crystals in distilled water that has been acidified with a few drops of concentrated hydrochloric acid and heated to 60 to 65°C. Take 4 ml for each gram of salt. Immediately filter the solution through a fritted-glass Hirsch funnel *into* ice-cold concentrated hydrochloric acid (1.5 ml for each gram of salt) in the vacuum filter flask. Transfer this filtrate to a small beaker and cool to −5°C in an ice–sodium chloride mixture. Red violet crystals should form; addition of a seed crystal may facilitate crystallization. Collect the crystals on the Hirsch funnel, wash well with ethanol, and then wash with ether. **(Caution! Keep ethanol and ether away from flames and sparks!)** Use the Teflon policeman to remove the last traces of product from the Hirsch funnel; scraping with a metal spatula may scour a cavity in the fritted-glass disc. Dry well in air, and store in a desiccator.

Resolution of *l-cis*-[Cr en$_2$Cl$_2$]Cl from the *d-l* Mixture.

Dissolve 1 gram of *cis*-[Cr en$_2$Cl$_2$]Cl·H$_2$O in 17 ml of water at room temperature. Add 1.9 grams of ammonium-*d*-α-bromocamphor-π-sulfonate with vigorous stirring. Keep the solution at room temperature until crystals form. Collect the crystals on the Hirsch funnel, and wash them with ethanol and then ether. The product at this stage is a salt, the cation being *l-cis*-[Cr en$_2$Cl$_2$]$^+$, the anion being *d*-α-bromocamphor-π-sulfonate.

Dissolve the crystals in concentrated hydrochloric acid, taking 4 ml of acid for each gram of salt. Add 2 volumes of ethanol for each volume of solution, and collect the resulting red-violet crystals on the Hirsch funnel. These crystals are the desired chloride salt of *l-cis*-[Cr en$_2$Cl$_2$]$^+$ cation. Wash with ethanol and ether. Dry well in air, and store in a desiccator.

Alternatively, the compound *d-cis*-[Cr en$_2$Cl$_2$]Cl can be resolved from the *d-l* mixture by using ammonium-*l*-α-bromocamphor-π-sulfonate.

BIBLIOGRAPHICAL NOTE

An excellent treatment of the development of theories about optical rotation and a description of experimental methods used in measuring optical rotation is T. M. Lowry, *Optical Rotatory Power*, London: Longmans, Green & Co., Ltd., 1935; paperback reprint by Dover Publications, Inc., New York, 1964. A more recent treatment of theory is J. A. Schellman, *The Optical Rotatory Properties of Proteins and Polypeptides*, Copenhagen: Danish Science Press, Ltd., 1958. A short section of theory and much emphasis on experimental practice is found in W. A. Struck and E. C. Olson, "Optical Rotation: Polarimetry," in I. M. Kolthoff, P. J. Elving, and E. B. Sandell, eds., *Treatise on Analytical Chemistry*, part I, vol. 6, New York: John Wiley & Sons, Inc., 1965. A fine book with emphasis on applications in organic chemistry is C. Djerassi, *Optical Rotatory Dispersion*, New York: McGraw-Hill, Inc., 1960; chapter 12, "Theory and Analysis of Rotatory Dispersion Curves" by A. Moscowitz, is especially recommended. An excellent introduction to polarization of light and to the nature of polarized light is the paperback, W. A. Shurcliff and S. S. Ballard, *Polarized Light*, New York: Van Nostrand Reinhold Company, 1964. There is a great deal of general information about polarimetry, as well as details regarding application of polarimetry to the chemistry of carbohydrates and to the technology of the sugar industry, in F. J. Bates et al., *Polarimetry, Saccharimetry and the Sugars*, Washington, D.C.: U.S. Government Printing Office, 1942 (National Bureau of Standards Circular C440).

The bromopentaaquochromium(III) cation can be prepared readily in the student laboratory, and the rate of its aquation can be followed spectrophotometrically. A student experiment, involving both preparation of the complex ion and a kinetic run, has been described by I. J. Herman and A. Lifshitz, *J. Chem. Educ.*, **47,** 231 (1970).

5

Chemical Relaxation

Three kineticists shared the 1967 Nobel Prize in chemistry. Dr. Manfred Eigen of the Max Planck Institute for Physical Chemistry in Göttingen, West Germany, together with Dr. Ronald G. W. Norrish and Dr. George Porter, Cambridge University, England, were recognized for their pioneering research in studying the rates and mechanisms of chemical reactions using relaxation techniques. The development of a wide variety of relaxation techniques opened a hitherto inaccessible time range with the measurement of rates of chemical reactions with half-lives of less than 1 nanosecond (10^{-9} second). These experiments eliminated the phrase "instantaneous chemical reaction" from chemistry books, for relaxation rates were measured for what are probably the fastest of all chemical reactions.

What are these powerful relaxation methods? The question will be answered in several steps. In this chapter the concepts of chemical relaxation processes and of chemical relaxation spectra will be studied for the cases of some unimolecular coupled reactions. Then later, in Chapter 8, the use of first-order relaxation methods for investigation of bimolecular mechanisms will be examined. And it will be shown how experiments can be designed so that the rate of most complex reactions can be observed as a spectrum of first-order relaxation processes.

We have, in fact, already encountered chemical relaxation. The exponential time–property curve associated with the unimolecular mechanism (3.3) described the way an initially nonequilibrium chemical system relaxes toward equilibrium. When the mechanism is more complicated, the chemical system relaxes from the nonequilibrium state toward equilibrium via several simultaneous relaxation processes. In this chapter, mechanisms will be examined involving two or more coupled unimolecular elementary reaction steps. The fundamental principles will be developed for the simplest systems to which they apply in order to give insight into the structure of first-order relaxation processes. The more complicated cases involving bimolecular elementary steps will be saved for Chapter 8, where several situations will be investigated in which the approach to equilibrium by bimolecular reactions occurs via first-order relaxation processes.

Sometimes there is clear experimental evidence for more than one elementary reaction step in the chemical mechanism. For example, the optical rotation data presented in Problem 5.6 cannot be represented by an equation with a single exponential term; the data can be represented by an equation with two exponential terms. This experimental rate curve is described as a superposition of two relaxations. A mechanism with two elementary reactions gives rise to a predicted rate curve of optical rotation versus time which is consistent with the experimentally observed rate curve. We shall now examine such a mechanism, paying particular attention to the procedure whereby we are able to move from the chemical statement of the mechanism to a mathematical equation for the rate curve.

Two Coupled Unimolecular Reactions

The chemical conversion of A into C described by the stoichiometric chemical equation

$$A = C \qquad (5.1)$$

may proceed through the formation of a chemical species of intermediate structure. A mechanism that includes such an intermediate is

$$A \underset{k_{-1}}{\overset{k_1}{\rightleftharpoons}} B \qquad (5.2)$$

$$B \underset{k_{-2}}{\overset{k_2}{\rightleftharpoons}} C \qquad (5.3)$$

This chemical reaction mechanism is a composite of two distinct microscopic reactions. Although this mechanism is often classified as a consecutive unimolecular reaction mechanism, it is important to recognize that (5.2) and (5.3) are simultaneous reactions which will be taking place whenever there are reactants or products available for reaction. The overall reaction is the superposition in time of the two elementary reactions. These two elementary reactions will be considered as independent reactions except for the coupling which occurs due to the fact that the concentration (B) which enters into the mathematical equations describing (5.2) is precisely the same quantity as the concentration (B) which enters into the corresponding mathematical equations describing (5.3).

The mathematical statement of this mechanism is the set of three simultaneous differential equations:

$$\frac{d(A)}{dt} = \left(\begin{array}{c} -k_1(A) \\ \textit{destruction of A} \end{array}\right) + \left(\begin{array}{c} k_{-1}(B) \\ \textit{formation of A} \end{array}\right) \tag{5.4}$$

$$\frac{d(B)}{dt} = \left(\begin{array}{c} -k_{-1}(B) - k_2(B) \\ \textit{destruction of B} \end{array}\right) + \left(\begin{array}{c} k_1(A) + k_{-2}(C) \\ \textit{formation of B} \end{array}\right) \tag{5.5}$$

$$\frac{d(C)}{dt} = \left(\begin{array}{c} -k_{-2}(C) \\ \textit{destruction of C} \end{array}\right) + \left(\begin{array}{c} k_2(B) \\ \textit{formation of C} \end{array}\right) \tag{5.6}$$

The desired solutions of Equations (5.4), (5.5), and (5.6) are equations for the time dependence of concentrations (A), (B), and (C). Trial solutions are assumed of the form

$$(A) = ae^{-mt} \qquad (B) = be^{-mt} \qquad (C) = ce^{-mt} \tag{5.7}$$

These trial solutions will be substituted into the three differential rate equations. In order to perform the substitution, three time derivatives[1] are also needed. Differentiation of each of Equations (5.7) gives these needed derivatives:

$$\frac{d(A)}{dt} = -ame^{-mt} \qquad \frac{d(B)}{dt} = -bme^{-mt} \qquad \frac{d(C)}{dt} = -cme^{-mt} \tag{5.8}$$

Substitution of Equations (5.7) and (5.8) into the three simultaneous differential rate equations (5.4), (5.5), and (5.6) yields

$$-ame^{-mt} = -k_1ae^{-mt} + k_{-1}be^{-mt} \tag{5.9}$$

$$-bme^{-mt} = -k_{-1}be^{-mt} - k_2be^{-mt} + k_1ae^{-mt} + k_{-2}ce^{-mt} \tag{5.10}$$

$$-cme^{-mt} = -k_{-2}ce^{-mt} + k_2be^{-mt} \tag{5.11}$$

---

[1] See Appendix II for a discussion of the derivative of the exponential function.

Division of both sides of each equation by the common exponential factor, followed by slight rearrangement, gives

$$a[k_1 - m] - b[k_{-1}] \qquad\qquad\qquad\qquad = 0 \qquad (5.12)$$

$$-a[k_1] \qquad + b[k_{-1} + k_2 - m] - c[k_{-2}] \qquad = 0 \qquad (5.13)$$

$$- b[k_2] \qquad\qquad\qquad + c[k_{-2} - m] = 0 \qquad (5.14)$$

It may be anticipated by analogy with Equations (3.28)–(3.31) that the quantities a, b, and c will turn out to be functions of initial concentrations of reactants. However, m and the k's will be expected to be constants at a particular temperature and pressure. Thus we inquire into the restraints that must be imposed on the constant coefficients in brackets in Equations (5.12)–(5.14) so that a, b, and c can take on arbitrary values. Said in somewhat different terms, there is a range of values of a within which a is a continuous variable under experimental control of the chemist, and a must be treated as an independent variable. The same considerations apply to b and to c. We proceed by eliminating the three independent variables among the three simultaneous equations (5.12), (5.13), and (5.14).

There are two equivalent methods of eliminating the independent variables. These two methods are closely matched in difficulty and in time consumption for this particular problem. But for a slightly more complicated problem, the method of determinants saves a great deal of time and frustration.

Brute-Force Method. Equation (5.12) is solved for a, Equation (5.14) is solved for c, and there results

$$a = \frac{bk_{-1}}{k_1 - m} \qquad c = \frac{bk_2}{k_{-2} - m} \qquad\qquad (5.15)$$

Equations (5.15) can be substituted into Equation (5.13) to give

$$-\frac{bk_1k_{-1}}{k_1 - m} + b[k_{-1} + k_2 - m] - \frac{bk_2k_{-2}}{k_{-2} - m} = 0 \qquad (5.16)$$

Division by b and multiplication by the factors $[k_1 - m]$ and $[k_{-2} - m]$ yields

$$-k_1k_{-1}[k_{-2} - m] + [k_{-1} + k_2 - m][k_1 - m][k_{-2} - m]$$

$$-k_2k_{-2}[k_1 - m] = 0 \qquad (5.17)$$

Expansion of Equation (5.17) results in 16 terms, 8 of which cancel, leaving the cubic equation in m:

$$m^3 - m^2[k_1 + k_{-1} + k_2 + k_{-2}] + m[k_{-1}k_{-2} + k_1k_2 + k_1k_{-2}] = 0 \quad (5.18)$$

Method of Determinants.   Equations (5.12), (5.13), and (5.14) are written in a form well suited for the use of determinants[2] in obtaining a solution relating m to the four k's. The solution is the determinantal equation obtained by equating the three-by-three determinant of the coefficients of a, b, and c to zero. Thus

$$\begin{vmatrix} [k_1 - m] & -[k_{-1}] & 0 \\ -[k_1] & [k_{-1} + k_2 - m] & -[k_{-2}] \\ 0 & -[k_2] & [k_{-2} - m] \end{vmatrix} = 0 \quad (5.19)$$

Expansion of the determinant yields the equation

$$[k_1 - m]\begin{vmatrix} [k_{-1} + k_2 - m] & -[k_{-2}] \\ -[k_2] & [k_{-2} - m] \end{vmatrix}$$

$$+ [k_{-1}]\begin{vmatrix} -[k_1] & -[k_{-2}] \\ 0 & [k_{-2} - m] \end{vmatrix} = 0 \quad (5.20)$$

The two-by-two determinants of Equation (5.20) can each be expanded to give the algebraic equation

$$[k_1 - m][k_{-1} + k_2 - m][k_{-2} - m]$$

$$- [k_1 - m][k_{-2}][k_2] - [k_{-1}][k_1][k_{-2} - m] = 0 \quad (5.21)$$

Equation (5.21) is identical with Equation (5.17), and so performing the indicated multiplication of polynomials must give Equation (5.18).

Macroscopic and Microscopic Rate Constants.   Three different values of m will satisfy Equation (5.18). These three values of m are the three roots

---

[2] See Appendix I for a discussion of the use of determinants for the solution of simultaneous linear equations.

of this cubic equation. One way to find these roots is to try to write the equation in the form

$$[m - m_0][m - m_1][m - m_2] = 0 \tag{5.22}$$

When the indicated multiplications are performed, Equation (5.22) becomes

$$m^3 - m^2[m_0 + m_1 + m_2] + m[m_1m_2 + m_0m_2 + m_0m_1] - m_0m_1m_2 = 0 \tag{5.23}$$

There are three roots, or three zeros, of Equation (5.22), because the equation will be satisfied when any one of the three bracketed factors is zero. These three roots are

$$m = m_0 \qquad m = m_1 \qquad m = m_2 \tag{5.24}$$

Now compare Equations (5.18) and (5.23). These two equations must be simultaneously valid for three different values of m. This means that the coefficient of each power of m must be the same for both equations. The sought-for relationship between the m's and the k's can be found by equating coefficients of like powers of m in the two equations. There result three equations:

$$m_0 + m_1 + m_2 = k_1 + k_{-1} + k_2 + k_{-2} \tag{5.25}$$

$$m_1m_2 + m_0m_2 + m_0m_1 = k_{-1}k_{-2} + k_1k_2 + k_1k_{-2} \tag{5.26}$$

$$m_0m_1m_2 = 0 \tag{5.27}$$

Equation (5.27) requires that one of the roots must be zero, and we pick m_0. The value $m_0 = 0$ is then substituted back into Equations (5.25) and (5.26), with the results

$$m_0 = 0 \tag{5.28}$$

$$m_1 + m_2 = k_1 + k_{-1} + k_2 + k_{-2} \tag{5.29}$$

$$m_1m_2 = k_{-1}k_{-2} + k_1k_2 + k_1k_{-2} \tag{5.30}$$

The quantities m_1 and m_2 are the two experimentally observable, nonzero macroscopic rate constants of the reaction. The four k's are the microscopic rate constants of the mechanism. There are four k's and just two m's, so it is not possible to evaluate any microscopic constant solely from information about the numerical values of the m's.

Alternatively, Equation (5.18) can be divided by m. This division cannot be performed if $m = 0$, so the results will apply only to the two nonzero m's. A quadratic equation results:

$$m^2 - m[k_1 + k_{-1} + k_2 + k_{-2}] + k_{-1}k_{-2} + k_1k_2 + k_1k_{-2} = 0 \quad (5.31)$$

The quadratic formula gives directly

$$m_1 = \frac{k_1+k_{-1}+k_2+k_{-2}+\sqrt{(k_1+k_{-1}+k_2+k_{-2})^2-4(k_{-1}k_{-2}+k_1k_2+k_1k_{-2})}}{2}$$

$$(5.32)$$

$$m_2 = \frac{k_1+k_{-1}+k_2+k_{-2}-\sqrt{(k_1+k_{-1}+k_2+k_{-2})^2-4(k_{-1}k_{-2}+k_1k_2+k_1k_{-2})}}{2}$$

$$(5.33)$$

Equations (5.32) and (5.33) are completely consistent with Equations (5.29) and (5.30).

Our next step in developing an understanding of the rate equations for this reaction mechanism involves an examination of certain special cases in which Equations (5.29)–(5.33) simplify. We shall look first at the conditions under which each m can be written as a sum of k's. We shall then examine the simplification which occurs when experiments yield m's that differ greatly in numerical values. Finally, we shall look at certain other limiting cases of special chemical interest.

Macroscopic Rate Constants as Sums of Microscopic Rate Constants. There are situations in which an individual macroscopic m can be identified with just one or two elementary processes, with just one or two k's. Such a situation would result if the two m's were sums of k's with each k contributing significantly to the value of just one of the m's.

There are seven ways that four different k's can be separated into two sets. We shall look at each of these possible groupings and see what restrictions would have to be imposed on the values of the individual rate constants so that both Equations (5.29) and (5.30) are simultaneously satisfied by solutions of the form

$$m_1 = k_i + k_j + \cdots$$

$$m_2 = k_m + k_n + \cdots$$

where the sets $\{k_i, k_j, \ldots\}$ and $\{k_m, k_n, \ldots\}$ have no elements in common.

The restrictions on the rate constants will be described in terms of inequalities. The inequality symbol $\gg$ is read "much greater than," and the symbol $\ll$ is read "much less than." These two symbols are used to indicate that the smaller quantity is so much smaller that it can be disregarded when computing the sum of the two quantities. Thus the inequality

$$\xi \gg \zeta$$

requires that

$$\xi + \zeta \simeq \xi$$

The symbol $\simeq$ is read "approximately equal to" and will be used when we wish to make no numerical distinction between the two sides of the equation.

1. Let $m_1 \simeq k_1 + k_{-1}$ and $m_2 \simeq k_2 + k_{-2}$. Equation (5.29) is satisfied. Multiplication of the two approximate m's gives

$$m_1 m_2 \simeq k_{-1}k_{-2} + k_1 k_2 + k_1 k_{-2} + k_{-1}k_2 \tag{5.34}$$

Equations (5.30) and (5.34) are equivalent if the term $k_{-1}k_2$ is so small that it can be neglected with respect to one or more of the other terms of the equation. This could occur if

$$k_{-1}k_2 \ll k_{-1}k_{-2}$$

and this inequality would be valid if

$$k_2 \ll k_{-2} \tag{5.35}$$

Inequality (5.35) allows us to write

$$m_2 \simeq k_{-2} \tag{5.36}$$

Alternatively, Equations (5.30) and (5.34) become equivalent if

$$k_{-1}k_2 \ll k_1 k_2$$

which in turn would be valid if

$$k_{-1} \ll k_1 \tag{5.37}$$

Inequality (5.37) allows us to write

$$m_1 \simeq k_1 \tag{5.38}$$

2. Let $m_1 \simeq k_1 + k_2$ and $m_2 \simeq k_{-1} + k_{-2}$. Equation (5.29) is satisfied.

The two approximate m's are multiplied together, giving

$$m_1 m_2 \simeq k_1 k_{-1} + k_2 k_{-1} + k_2 k_{-2} + k_1 k_{-2} \tag{5.39}$$

Equations (5.30) and (5.39) have but one term in common. In order for the two equations to be equivalent, all other terms in both equations must be insignificantly small compared to $k_1 k_{-2}$. A sufficient set of inequalities is

$$k_2 \ll k_{-2}$$
$$k_2 \ll k_1$$
$$k_{-1} \ll k_{-2} \tag{5.40}$$
$$k_{-1} \ll k_1$$

There then is obtained

$$m_1 \simeq k_1$$
$$m_2 \simeq k_{-2} \tag{5.41}$$

Inequalities (5.40) are thus seen to be sufficient to decouple the two elementary reactions (5.2) and (5.3) in the sense that macroscopic rate constant m_1 is a function of the microscopic rate constants for elementary reaction (5.2) only, and that similarly m_2 is a function of the k's for just elementary reaction (5.3).

3. Let $m_1 \simeq k_1 + k_{-2}$ and $m_2 \simeq k_{-1} + k_2$. Equation (5.29) is satisfied. The two approximate m's are multiplied together to give

$$m_1 m_2 \simeq k_1 k_{-1} + k_1 k_2 + k_{-1} k_{-2} + k_{-2} k_2 \tag{5.42}$$

Conditions must be imposed which assure that terms $k_1 k_{-1}$, $k_{-2} k_2$, and $k_1 k_{-2}$ may be neglected as small with respect to one or more of the other terms. This can be accomplished by having

$$k_1 \ll k_{-2} \quad \text{or} \quad k_{-1} \ll k_2$$
$$k_1 \gg k_{-2} \quad \text{or} \quad k_{-1} \gg k_2$$
$$k_1 \ll k_{-1}$$
$$k_{-2} \ll k_2$$

Thus, if

$$k_1 \ll k_{-2} \ll k_2 \ll k_{-1}$$

then

$$m_1 \simeq k_{-2} \quad \text{and} \quad m_2 \simeq k_{-1} \tag{5.43}$$

Or, if

$$k_{-2} \ll k_1 \ll k_{-1} \ll k_2$$

then

$$m_1 \simeq k_1 \quad \text{and} \quad m_2 \simeq k_2 \tag{5.44}$$

These are stringent requirements, compared to the previous two cases. Decoupling of the two elementary reactions results.

4. Let $m_1 \simeq k_1$ and $m_2 \simeq k_{-1} + k_2 + k_{-2}$. Equation (5.29) is satisfied. The two approximate m's are multiplied together, giving

$$m_1 m_2 \simeq k_1 k_{-1} + k_1 k_2 + k_1 k_{-2} \tag{5.45}$$

Comparison of Equations (5.30) and (5.45) shows that for the two to be equivalent, it is necessary to have terms $k_1 k_{-1}$ and $k_{-1} k_{-2}$ insignificantly small. For this to occur, it is sufficient that

$$k_{-1} \ll k_1$$

$$k_{-1} \ll k_2 \quad \text{or} \quad k_{-1} \ll k_{-2}$$

The requirement here is essentially that k_{-1} be small. Either k_2 or k_{-2}, but not both, can also be small. Macroscopic rate constant m_2 is then given by

$$m_2 \simeq k_2 + k_{-2} \tag{5.46}$$

5. Let $m_1 \simeq k_{-1}$ and $m_2 \simeq k_1 + k_2 + k_{-2}$. Equation (5.29) is satisfied. Multiplication of the two approximate m's yields

$$m_1 m_2 \simeq k_1 k_{-1} + k_{-1} k_2 + k_{-1} k_{-2} \tag{5.47}$$

Equations (5.30) and (5.47) have but one term in common. For the two equations to be equivalent, it is sufficient that

$$k_2 \ll k_{-2}$$

$$k_1 \ll k_{-2}$$

$$k_1 \ll k_{-1}$$

These inequalities yield

$$m_2 \simeq k_{-2} \tag{5.48}$$

6. The remaining two possibilities yield nothing new, because of the symmetry of the mechanism.

We have just seen ways in which it is possible to make assumptions about the relative values of the microscopic rate constants of the mechanism in order that each m will be equal to a sum of k's.

In each case the assumptions were sufficient to decouple the two elementary reactions (5.2) and (5.3). Each m became a function of the k's of just one elementary reaction.

The assumptions made become integral parts of the assumed mechanism. We have not discussed any methods whereby such assumptions can be directly tested by experiment.

Simplification When Macroscopic Rate Constants Differ Greatly in Value.
If it is found experimentally that the two macroscopic rate constants have quite different numerical values, then the m's can be written individually as explicit functions of the k's. Equation (5.29) becomes

$$m_1 + m_2 \simeq m_1 \simeq k_1 + k_{-1} + k_2 + k_{-2} \tag{5.49}$$

where it has been arbitrarily decided that m_1 is the larger m. Division of Equation (5.30) by (5.29) gives

$$\frac{m_1 m_2}{m_1 + m_2} \simeq \frac{m_1 m_2}{m_1} = m_2 \simeq \frac{k_{-1}k_{-2} + k_1 k_2 + k_1 k_{-2}}{k_1 + k_{-1} + k_2 + k_{-2}} \tag{5.50}$$

The approximation that $m_1 + m_2 \simeq m_1$ is not an arbitrary assumption. Its validity depends directly on the results of kinetics experiments. The approximation is valid whenever the observed values of the two macroscopic rate constants differ greatly in magnitude.

Chemical Interpretation of Some Limiting Cases. There is a great deal of chemistry hidden in Equations (5.29) and (5.30). We shall try to find some of this hidden meaning by examining in turn a mechanism with a fast elementary reaction followed by a slow reaction, another mechanism in which the intermediate B can be considered in a steady state during the second relaxation process, and finally a third mechanism with two virtually irreversible elementary reactions.

1. $m_1 \gg m_2$, k_1 and k_{-1} individually much greater than k_2 and k_{-2}. This set of inequalities corresponds to a reaction mechanism which can be written as

$$A \underset{k_{-1}}{\overset{k_1}{\rightleftharpoons}} B \qquad \textit{fast} \tag{5.51}$$

$$B \underset{k_{-2}}{\overset{k_2}{\rightleftharpoons}} C \qquad \textit{slow} \tag{5.52}$$

There are two quite distinct types of inequalities in this limiting case. The inequality $m_1 \gg m_2$ states the experimental fact. The assertion that the rate constants for Reaction (5.51) are individually much greater than the rate constants for Reaction (5.52) is an assumption, an integral part of the proposed mechanism. Equation (5.49) becomes

$$m_1 \simeq k_1 + k_{-1} \tag{5.53}$$

Equation (5.50) becomes

$$m_2 \simeq \frac{k_{-1}k_{-2} + k_2k_1 + k_1k_{-2}}{k_1 + k_{-1}} \tag{5.54}$$

It is easier to read chemical meaning into Equation (5.54) if both numerator and denominator of the fraction are divided by k_{-1} and if the equilibrium constant for reaction step (5.51) is then introduced by means of Equation (5.65). These operations result in

$$m_2 \simeq \frac{k_{-2} + K_1'[k_2 + k_{-2}]}{K_1' + 1} = \frac{[K_1' + 1]k_{-2} + K_1'k_2}{K_1' + 1}$$

$$= k_{-2} + \frac{K_1'}{K_1' + 1} k_2 \tag{5.55}$$

Equation (5.55) becomes even simpler if the value of K_1' is quite different from unity. Thus

$$K_1' \gg 1: \quad m_2 \simeq \frac{K_1'k_{-2} + K_1'k_2}{K_1'} = k_{-2} + k_2 \tag{5.56}$$

$$K_1' \ll 1: \quad m_2 \simeq k_{-2} + K_1'k_2 \tag{5.57}$$

It often happens that the experimental methods being used to follow the reaction are too slow to give information about the rate of the fast relaxation characterized by macroscopic rate constant m_1; the observed rate of reaction is then just the first-order relaxation characterized by m_2. In this case, it is often said that the slow reaction (5.52) is the *rate-determining step* of the mechanism, because m_2 is a function of the rate constants for (5.52), with the possible inclusion of the equilibrium constant for the fast step (5.51). Of course, if the fast relaxation process had been observed, it would be perfectly reasonable to say that the fast reaction (5.51) was the rate-determining step for the fast relaxation.

The assumed inequalities have in effect decoupled the two elementary reactions. The fast relaxation can be identified with the fast elementary reaction. Although K_1' appears in the expression for m_2, neither k_1 nor k_{-1} individually does, and so the *rate* of the fast reaction does not influence the value of m_2. The slow relaxation proceeds at a rate governed by the slow elementary reaction, modified only by the inclusion of the equilibrium constant. This situation is sometimes described by saying that the slow relaxation is preceded by a "rapid preequilibrium step." Since the values of (A) and (B) continue to change throughout the slow relaxation, it is clear that this "preequilibrium" is not an equilibrium state in the usual sense.

2. $m_1 \gg m_2$, k_{-1} and k_2 individually much greater than k_1 and k_{-2}. Since these inequalities require that K_1' defined by Equation (5.65) be small and that K_2' defined by Equation (5.66) be large, this is a limiting case in which the equilibrium concentration of the intermediate B is very small compared to concentrations (A) and (C). This low relative concentration of the intermediate makes the mechanism interesting in a discussion of the steady state.[3] Again it is important to note that the inequality involving the m's must be verified directly by experiment, whereas the inequality involving the k's is assumed as part of the proposed mechanism. Equation (5.49) becomes

$$m_1 \simeq k_{-1} + k_2 \tag{5.58}$$

Equation (5.50) becomes

$$m_2 \simeq \frac{k_1 k_2 + k_{-1} k_{-2}}{k_{-1} + k_2} \tag{5.59}$$

Neither macroscopic rate constant can be identified with a single elementary step of the mechanism. Each observable relaxation is a composite of the two elementary reactions. These elementary reactions are strongly coupled, superimposed reactions.

3. k_1 and k_2 individually much greater than k_{-1} and k_{-2}. This mechanism consists of two virtually irreversible elementary reactions. At equilibrium essentially all the starting material will be present as compound C. Equations (5.29) and (5.30) become

$$m_1 + m_2 = k_1 + k_2 \tag{5.60}$$

$$m_1 m_2 \simeq k_1 k_2 \tag{5.61}$$

A pair of solutions[4] to Equations (5.60) and (5.61) is

$$m_1 \simeq k_1$$
$$m_2 \simeq k_2 \tag{5.62}$$

[3] See page 159.
[4] Another pair of solutions is $m_1 \simeq k_2$, $m_2 \simeq k_1$. Are there any other solutions?

If k_1 and k_2 are of about the same numerical value, then only one relaxation will be observed. The observed rate then becomes experimentally indistinguishable from the rate predicted for Mechanism (3.3) by Equations (3.32) and (3.33).

Rate Constants and Equilibrium Constants. After equilibrium has been achieved, there is no further net change with time of any of the concentrations, even though both elementary reactions are continuing to interconvert species A, B, and C. At equilibrium the three derivatives $d(A)/dt$, $d(B)/dt$, and $d(C)/dt$ are individually equal to zero, and Equations (5.4) and (5.6) become

$$k_1(\overline{A}) = k_{-1}(\overline{B}) \tag{5.63}$$

$$k_{-2}(\overline{C}) = k_2(\overline{B}) \tag{5.64}$$

Equations (5.63) and (5.64) can be used to state equilibrium-constant relationships for Reactions (5.2) and (5.3):

$$\frac{(\overline{B})}{(\overline{A})} = \frac{k_1}{k_{-1}} \equiv K_1' \tag{5.65}$$

$$\frac{(\overline{C})}{(\overline{B})} = \frac{k_2}{k_{-2}} \equiv K_2' \tag{5.66}$$

The overall equilibrium constant for stoichiometric reaction (5.1) can be written in terms of the equilibrium constants for the two elementary reaction steps by multiplying together K_1' and K_2':

$$\frac{(\overline{B})}{(\overline{A})} \frac{(\overline{C})}{(\overline{B})} = \frac{(\overline{C})}{(\overline{A})} = K_1' K_2' \equiv K' \tag{5.67}$$

This arithmetic can proceed in only one direction. If both rate constants for an elementary reaction are known, then the equilibrium constant for that reaction can be calculated; the knowledge of the numerical value of the equilibrium constant, however, gives no information by itself about the individual values of the two rate constants. Similarly, if the value of each of the equilibrium constants for individual reactions is known, then the overall equilibrium constant can be calculated; knowledge of the value of the overall equilibrium constant gives no information about the individual values of the equilibrium constants for the elementary reaction steps.

Relaxation Spectrum.   The three roots m_0, m_1, and m_2 of Equation (5.18) generate three particular solutions which correspond to each of the trial solutions (5.7). There result nine such particular solutions. No one of these particular solutions is valid by itself, and the three particular solutions cannot be simultaneously valid. The chemically meaningful solution is the general solution obtained by forming the sum of the three particular solutions for each of the trial equations (5.7). Thus the chemically meaningful solution is the set of three simultaneous equations:

$$(A) = a_0 e^{-m_0 t} + a_1 e^{-m_1 t} + a_2 e^{-m_2 t} \tag{5.68}$$

$$(B) = b_0 e^{-m_0 t} + b_1 e^{-m_1 t} + b_2 e^{-m_2 t} \tag{5.69}$$

$$(C) = c_0 e^{-m_0 t} + c_1 e^{-m_1 t} + c_2 e^{-m_2 t} \tag{5.70}$$

Since $e^{-m_0 t} = e^{(0)} = 1$, Equations (5.68)–(5.70) become

$$(A) = a_0 + a_1 e^{-m_1 t} + a_2 e^{-m_2 t} \tag{5.71}$$

$$(B) = b_0 + b_1 e^{-m_1 t} + b_2 e^{-m_2 t} \tag{5.72}$$

$$(C) = c_0 + c_1 e^{-m_1 t} + c_2 e^{-m_2 t} \tag{5.73}$$

The experimental rate curve for this reaction, when first viewed as it is being traced out by the laboratory strip-chart recorder, may seem quite complex. After the composite rate curve has been resolved into contributions from several first-order relaxation processes, it is possible to analyze each relaxation process separately, and to determine for each relaxation the numerical value of its associated macroscopic first-order rate constant. This analyzed rate curve is interpreted as the superposition of a spectrum of individual relaxations spread out along a time axis. And the justification for this analysis of the complex curve as the sum of exponential terms is the set of simultaneous equations (5.71)–(5.73) which predict just such a rate curve.

Since the numerical values of the two m's and of the nine preexponential coefficients range over a fantastically wide span—depending on the particular reaction and on the experimental conditions—it is to be expected that the details of analyzing a relaxation spectrum will differ from case to case.

Properties of the Sum of Two Exponentials.   Insight into the effect on a function such as Equation (5.71) of varying the numerical values of the time-invariant constants can be gained by examining some calculated values for a slightly simpler function,

$$A = a_1 e^{-m_1 t} + a_2 e^{-m_2 t} \tag{5.74}$$

The results of some such calculations are presented in Figures 5.1 through 5.7. In each case the plot of A versus t is compared with a plot of $\ln A$ versus t for the same values of the a's and the m's. Equation (5.74) is, of course, just Equation (5.71) with a_0 set equal to zero, and with (A) replaced by an unnamed variable A which can be considered as a concentration or as absorbance or as any other property which seems convenient in making the following discussion more specific and less abstract.

Consider Figures 5.1, 5.2, and 5.3. The numerical values of the m's ($m_1 = 0.16$, $m_2 = 0.01$) remain the same throughout the series of three graphs. The ratio a_1/a_2 of the preexponential constants is varied from 9:1 to 1:1 to finally 1:9, and there is a marked effect on the shape of the resulting graphs. It is clear that a fast relaxation process can easily be masked if a_1 has a relatively small value, whereas the slow relaxation will generally be observable if the changes in the observed property are large enough to be measured by the experimental method being used.

The fast relaxation can be detected if changes in A due to the term $a_1 e^{-m_1 t}$ are large compared to changes in A due to the term $a_2 e^{-m_2 t}$ throughout a time interval of at least $1/m_1$ time units. Figure 5.3 shows the result of such large changes in A arising from the slow relaxation that the fast relaxation is barely detectable. Quite precise measurements of A would be required if a numerical value of m_1 were to be obtained from such a rate curve. In fact, quite precise measurements would be required to convince a sceptic that there were really two relaxation processes occurring. A much more desirable situation is seen in Figure 5.1, where the two relaxation processes are separable without difficulty.

Since A_∞ is equal to zero, the quantity $\ln |A_t - A_\infty|$ is equal to $\ln A$ at each value of t. Excluding the initial portion of the curve, then, the graph of $\ln A$ versus t is the infinite-time method of determining the macroscopic rate constant. If the graphs of A versus t given in the figures were, in fact, the complete available experimental data, the infinite-time value of A would not be known and it would be necessary to use the Guggenheim method or the time-lag method. Further discussion of ways to evaluate the two macroscopic rate constants from experimental data begins on page 85.

Consider Figures 5.4 through 5.7. This set of figures is presented to indicate the extent to which the numerical values of the m's must be different in order that the experimental rate curve will give clear evidence of two relaxations and in order that reliable determination of values of the two m's can be made. When the two m's have the same value, the resulting curve is a single relaxation curve. Comparison is made of plots of A versus t for m_1/m_2 ratios of 1:1, 4:1, 16:1, and 64:1, in each case with the two preexponential constants equal. For the 64-fold difference, the logarithmic plot consists of two virtually straight lines, joined by a curved section.

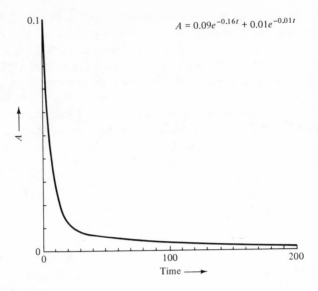

$$A = 0.09e^{-0.16t} + 0.01e^{-0.01t}$$

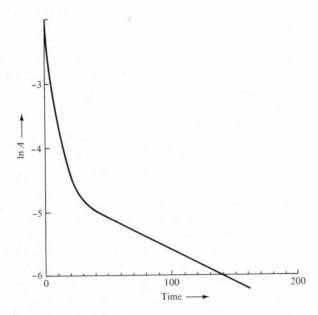

FIGURE 5.1. Sum of two exponentials, $a_1/a_2 = 9$, $m_1/m_2 = 16$

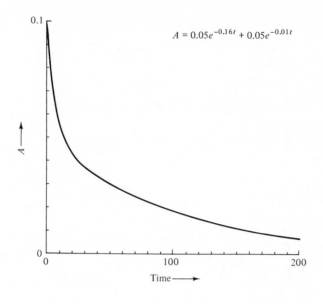

$$A = 0.05e^{-0.16t} + 0.05e^{-0.01t}$$

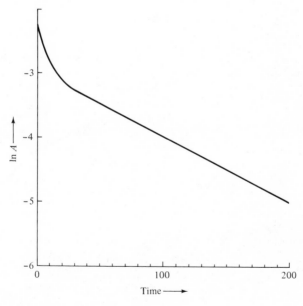

FIGURE 5.2.   Sum of two exponentials, $a_1/a_2 = 1$, $m_1/m_2 = 16$

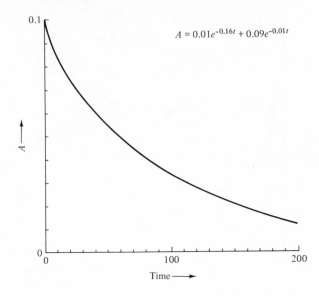

$$A = 0.01e^{-0.16t} + 0.09e^{-0.01t}$$

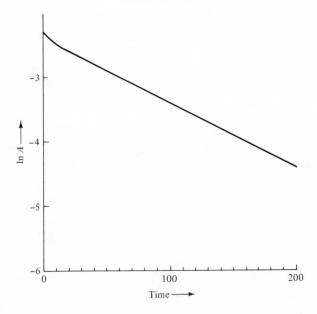

FIGURE 5.3. Sum of two exponentials, $a_1/a_2 = 1/9$, $m_1/m_2 = 16$

Examination of this collection of calculated rate curves leads one to the generalization that for ease in resolving the two relaxations from the experimental rate curve, it is desirable to have the combination of conditions

$$a_1 > 9a_2$$

$$m_1 > 16m_2$$

These are not absolute requirements. The greater the experimental uncertainties in the rate data, the greater the separation required in values of the a's and the m's.

In many experimental situations, the numerical values of the a's and of the m's are under the control of the chemist. For instance, if the reaction is being followed by absorbance measurements, a_1 and a_2 will in general be different functions of the absorptivities of the various chemical species. In turn, these absorptivities are ordinarily different functions of the wavelength of light being used for absorbance measurements. Thus it is often possible to find a particular wavelength of light which will yield a desired a_1/a_2 ratio. In Chapter 8 it is pointed out that most chemical systems which appear to react according to unimolecular elementary reaction steps actually have at least one bimolecular step. For all such cases, the macroscopic rate constants are concentration dependent. Typically, it is possible to find a concentration range in which the relaxation processes can be separated on different time scales, with the m's differing by several powers of 10.

Experimental Determination of Two Macroscopic Rate Constants

The greatest amount of reliable quantitative information can be extracted from a time–property curve with the least amount of effort if the numerical values of the two m's are different by several powers of 10, by several orders of magnitude. The two relaxations then occur on different time scales. It is possible to separate two such relaxations by choosing an appropriate time interval for each. We will restrict our attention to such convenient situations.[5]

[5] Mathematical methods have been developed to deal with situations where such convenient separation cannot be achieved. Some representative papers in which overlapping relaxations have been resolved include:

D. F. Abell, N. A. Bonner, and W. Goishi, *J. Chem. Phys.*, **27**, 658 (1957).

J.-P. Mathieu, *Bull. Soc. Chim. Fr.*, [5] **4**, 687 (1937).

J. Selbin and J. C. Bailar, Jr., *J. Amer. Chem. Soc.*, **79**, 4285 (1957).

G. Davies, K. Kustin, and R. F. Pasternack, *Int. J. Chem. Kinetics*, **1**, 45 (1969).

R. E. Smith and M. F. Morales, *Bull. Math. Biophys.*, **6**, 133 (1944).

When the values of m_1 and m_2 are nearly the same, very accurate rate data are needed to evaluate the two m's individually with reliability. Such very accurate experimental rate data are difficult to obtain for most chemically reacting systems.

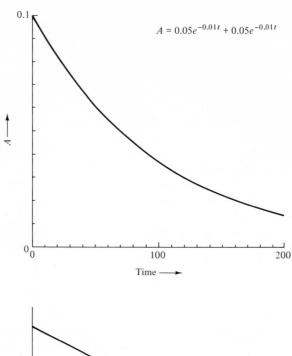

$$A = 0.05e^{-0.01t} + 0.05e^{-0.01t}$$

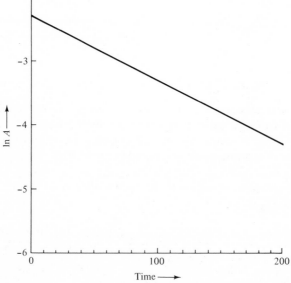

FIGURE 5.4. Sum of two exponentials, $a_1/a_2 = 1$, $m_1/m_2 = 1$

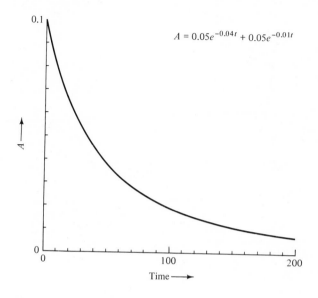

$$A = 0.05e^{-0.04t} + 0.05e^{-0.01t}$$

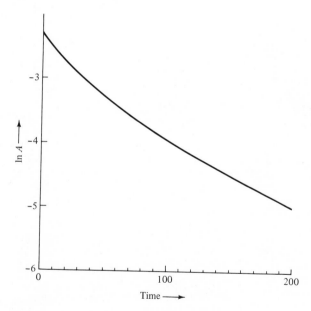

FIGURE 5.5.   Sum of two exponentials, $a_1/a_2 = 1$, $m_1/m_2 = 4$

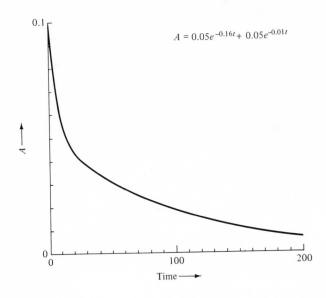

$$A = 0.05e^{-0.16t} + 0.05e^{-0.01t}$$

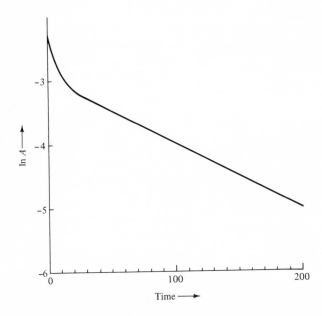

FIGURE 5.6. Sum of two exponentials, $a_1/a_2 = 1$, $m_1/m_2 = 16$

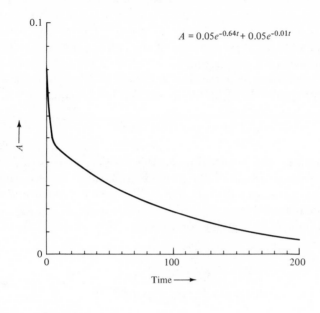

$$A = 0.05e^{-0.64t} + 0.05e^{-0.01t}$$

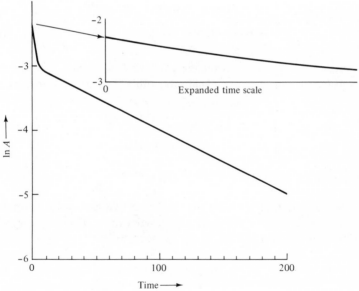

FIGURE 5.7. Sum of two exponentials, $a_1/a_2 = 1$, $m_1/m_2 = 64$

Use of optical rotation measurements to follow the reaction of A, B, and C will now be examined. The method being developed, although discussed here for polarimetry, is applicable to any property of the reacting system that is an additive linear function of the individual concentrations of reactants, intermediates, and products.

The optical rotation α of a solution is related to the concentrations of optically active species A, B, and C in solution by the equation

$$\alpha = l\alpha_A(A) + l\alpha_B(B) + l\alpha_C(C) \tag{5.75}$$

where l is the optical path length through the solution, and α_A, α_B, and α_C are constants characteristic of the individual chemical species, called the molar rotatory coefficients of A, B, and C. The time dependence of the optical rotation is obtained by substituting Equations (5.71), (5.72), and (5.73) into Equation (5.75), giving for Reaction Mechanism (5.2)–(5.3) the result

$$\begin{aligned}
\alpha_t &= l\{\alpha_A a_0 + \alpha_B b_0 + \alpha_C c_0\} \\
&\quad + l\{\alpha_A a_1 + \alpha_B b_1 + \alpha_C c_1\}e^{-m_1 t} \\
&\quad + l\{\alpha_A a_2 + \alpha_B b_2 + \alpha_C c_2\}e^{-m_2 t} \\
&= \mathcal{A}_0 + \mathcal{A}_1 e^{-m_1 t} + \mathcal{A}_2 e^{-m_2 t}
\end{aligned} \tag{5.76}$$

The quantity α_t is the instantaneous value of α at time t. The quantities $\mathcal{A}_0$, $\mathcal{A}_1$, and $\mathcal{A}_2$ are collections of constants, and are the time-invariant quantities

$$\mathcal{A}_0 = l\{\alpha_A a_0 + \alpha_B b_0 + \alpha_C c_0\}$$

$$\mathcal{A}_1 = l\{\alpha_A a_1 + \alpha_B b_1 + \alpha_C c_1\}$$

$$\mathcal{A}_2 = l\{\alpha_A a_2 + \alpha_B b_2 + \alpha_C c_2\}$$

We are considering the case in which the two first-order macroscopic rate constants have quite different numerical values. The arbitrary choice is made that of the two m's, m_1 is the larger. Thus

$$m_1 \gg m_2$$

When the time t is of the same order of magnitude as $1/m_1$, so that, for instance,

$$\frac{1}{10m_1} < t < \frac{10}{m_1} \tag{5.77}$$

then Equation (5.76) can be approximated by

$$\alpha_t \simeq \alpha_0 + \alpha_1 e^{-m_1 t} + \alpha_2 \qquad (5.78)$$

because the value of t is so small compared to $1/m_2$ that the second exponential factor is essentially equal to unity throughout the time interval (5.77). Provided that data are taken from within the appropriate time interval, Equation (5.78) can be treated just as Equation (3.35) was for evaluation of m_1 from experimental data. The Guggenheim method is ideally suited for making the logarithmic plot, and we shall now examine in detail how this method is applied.

The quantity $\alpha_{t+\tau}$ is defined by the expression

$$\alpha_{t+\tau} = \alpha \text{ at time } [t + \tau]$$

where τ is an arbitrary but constant interval of time chosen at the convenience of the experimenter. Equation (5.78) is written at t and also at time $[t + \tau]$, giving

$$\alpha_t = \alpha_0 + \alpha_2 + \alpha_1 e^{-m_1 t} \qquad (5.79)$$

$$\alpha_{t+\tau} = \alpha_0 + \alpha_2 + \alpha_1 e^{-m_1 [t+\tau]} \qquad (5.80)$$

The $\simeq$ sign has been replaced by the $=$ sign, so that there will be no implication that further approximations are being made. Subtraction of Equation (5.80) from Equation (5.79) leaves

$$\begin{aligned}
\alpha_t - \alpha_{t+\tau} &= \alpha_1 e^{-m_1 t} - \alpha_1 e^{-m_1 [t+\tau]} \\
&= \alpha_1 e^{-m_1 t} - \alpha_1 \{e^{-m_1 t}\} \{e^{-m_1 \tau}\} \\
&= \alpha_1 e^{-m_1 t} \{1 - e^{-m_1 \tau}\} \\
&= \{\text{constant}\} e^{-m_1 t} \qquad (5.81)
\end{aligned}$$

The logarithm to the base e is taken of both sides of Equation (5.81) to yield

$$\ln |\alpha_t - \alpha_{t+\tau}| = \text{constant} - m_1 t \qquad (5.82)$$

A plot of the logarithmic term in (5.82) versus t will be a straight line if Equation (5.78) is a good approximation of the actual reaction rate curve within the time interval in which the data were taken. The slope of the straight line is the negative of the experimental value of the macroscopic rate constant m_1.

In like manner, when t has a value comparable to $1/m_2$, we can approximate Equation (5.75) by

$$\alpha_t \simeq \alpha_0 + \alpha_2 e^{-m_2 t} \qquad (5.83)$$

because the first exponential factor is so small that the term can be neglected with respect to the other terms. Experimental evaluation of m_2 can proceed via the infinite-time, Guggenheim, time-lag, or half-life methods.

The infinite-time and half-life methods are often difficult to apply to evaluation of the macroscopic rate constant for the faster relaxation, because only in the cases of extremely different values of the two m's does the time–property curve level off to a pseudo infinite-time value on the time scale of the faster relaxation. The Guggenheim or time-lag method is usually preferred. If the time interval τ used in making the plot for faster relaxation is too long, a straight line may not be obtained. The Guggenheim and time-lag methods are not valid if the plot fails to be straight.

Cyclic Unimolecular Mechanisms

Reaction Mechanism (5.2)–(5.3) can be expanded to a cyclic mechanism without the addition of more intermediates. Let us add a third elementary reaction step, permitting direct interconversion of A and C. The complete mechanism is then

$$A \underset{k_{-1}}{\overset{k_1}{\rightleftarrows}} B \qquad (5.84)$$

$$B \underset{k_{-2}}{\overset{k_2}{\rightleftarrows}} C \qquad (5.85)$$

$$C \underset{k_{-3}}{\overset{k_3}{\rightleftarrows}} A \qquad (5.86)$$

The interconversion of any two of the species occurs by two simultaneous pathways: One pathway is the direct reaction, and the other pathway involves the third species as a reaction intermediate. Certain additional aspects of this mechanism are discussed in Chapter 6, beginning on page 106.

There are three simultaneous differential equations associated with the mechanism. Because of the cyclic nature of the mechanism, no species is unique, and the differential equations have a symmetry that was lacking in the set (5.4), (5.5), and (5.6). These differential equations, comprising the mathematical model for this mechanism, are

$$\frac{d(A)}{dt} = -k_1(A) - k_{-3}(A) + k_{-1}(B) + k_3(C) \tag{5.87}$$

$$\frac{d(B)}{dt} = -k_2(B) - k_{-1}(B) + k_{-2}(C) + k_1(A) \tag{5.88}$$

$$\frac{d(C)}{dt} = -k_3(C) - k_{-2}(C) + k_{-3}(A) + k_2(B) \tag{5.89}$$

Exponential trial solutions and their derivatives—Equations (5.7) and (5.8)—are substituted into the differential equations. Three equations analogous to Equations (5.9), (5.10), and (5.11) result:

$$-mae^{-mt} = -k_1ae^{-mt} - k_{-3}ae^{-mt} + k_{-1}be^{-mt} + k_3ce^{-mt}$$

$$-mbe^{-mt} = -k_2be^{-mt} - k_{-1}be^{-mt} + k_{-2}ce^{-mt} + k_1ae^{-mt}$$

$$-mce^{-mt} = -k_3ce^{-mt} - k_{-2}ce^{-mt} + k_{-3}ae^{-mt} + k_2be^{-mt}$$

Each term has the factor e^{-mt}. This common exponential factor is canceled, leaving three simultaneous algebraic equations that have no dependence on time. Slightly rearranged, these equations are

$$a[k_1 + k_{-3} - m] - b[k_{-1}] \qquad\qquad - c[k_3] \qquad\qquad = 0$$

$$-a[k_1] \qquad\qquad + b[k_2 + k_{-1} - m] - c[k_{-2}] \qquad\qquad = 0$$

$$-a[k_{-3}] \qquad\qquad - b[k_2] \qquad\qquad + c[k_3 + k_{-2} - m] = 0$$

We shall use the method of determinants to find solutions for these simultaneous equations. The determinant of the coefficients of a, b, and c is formed. Then the solution of the three simultaneous equations is the solution of the determinantal equation

$$\begin{vmatrix} [k_1 + k_{-3} - m] & -[k_{-1}] & -[k_3] \\ -[k_1] & [k_2 + k_{-1} - m] & -[k_{-2}] \\ -[k_{-3}] & -[k_2] & [k_3 + k_{-2} - m] \end{vmatrix} = 0 \tag{5.90}$$

Expansion of this three-by-three determinant gives three two-by-two determinants, and the determinantal equation becomes

$$
[k_1 + k_{-3} - m]
\begin{vmatrix}
[k_2 + k_{-1} - m] & -[k_{-2}] \\
-[k_2] & [k_3 + k_{-2} - m]
\end{vmatrix}
$$

$$
+ k_{-1}
\begin{vmatrix}
-[k_1] & -[k_{-2}] \\
-[k_{-3}] & [k_3 + k_{-2} - m]
\end{vmatrix}
$$

$$
- k_3
\begin{vmatrix}
-[k_1] & [k_2 + k_{-1} - m] \\
-[k_{-3}] & -[k_2]
\end{vmatrix} = 0 \quad (5.91)
$$

Further expansion of determinants yields the algebraic equation

$$
[k_1 + k_{-3} - m][k_2 + k_{-1} - m][k_3 + k_{-2} - m]
$$

$$
- k_2 k_{-2}[k_1 + k_{-3} - m] - k_1 k_{-1}[k_3 + k_{-2} - m]
$$

$$
- k_{-1} k_{-2} k_{-3} - k_1 k_2 k_3 - k_3 k_{-3}[k_2 + k_{-1} - m] = 0 \quad (5.92)
$$

When the indicated multiplications are carried out, it is found that several terms cancel. The terms that remain, collected by powers of m, give the cubic equation

$$
m^3 - m^2[k_1 + k_{-1} + k_2 + k_{-2} + k_3 + k_{-3}]
$$

$$
+ m[k_{-1} k_{-2} + k_{-1} k_3 + k_2 k_3 + k_{-2} k_{-3} + k_{-3} k_2
$$

$$
+ k_{-3} k_{-1} + k_1 k_2 + k_1 k_{-2} + k_1 k_3] = 0 \quad (5.93)
$$

The three roots of Equation (5.93) are found by equating the coefficients of like powers of m in (5.93) and (5.23). There result the three equations

$$
m_0 = 0 \quad (5.94)
$$

$$
m_1 + m_2 = k_1 + k_{-1} + k_2 + k_{-2} + k_3 + k_{-3} \quad (5.95)
$$

$$
m_1 m_2 = k_{-1} k_{-2} + k_{-1} k_3 + k_2 k_3 + k_{-2} k_{-3} + k_{-3} k_2
$$

$$
+ k_{-3} k_{-1} + k_1 k_2 + k_1 k_{-2} + k_1 k_3 \quad (5.96)
$$

Substitution of the three values of m into the exponential trial solutions results in nine particular solutions. None of these particular solutions is valid by itself. The chemically meaningful general solutions, formed by appropriate linear combinations of particular solutions, are

$$(A) = a_0 + a_1 e^{-m_1 t} + a_2 e^{-m_2 t} \tag{5.97}$$

$$(B) = b_0 + b_1 e^{-m_1 t} + b_2 e^{-m_2 t} \tag{5.98}$$

$$(C) = c_0 + c_1 e^{-m_1 t} + c_2 e^{-m_2 t} \tag{5.99}$$

Equations (5.97), (5.98,) and (5.99) have exactly the same form as Equations (5.71), (5.72), and (5.73). Thus there is no difference between the functional form of the time dependence of concentrations predicted for Mechanism (5.2)–(5.3) and that predicted for Mechanism (5.84)–(5.86). The meaning of the microscopic rate constants depends on the mechanism, and it is different in the two cases. The addition of the third pair of reaction arrows adds two more microscopic rate constants; the number of observable relaxation processes and consequently the number of macroscopic rate constants remains equal to two.

Time Course of the Individual Concentrations

Contained within the equations already developed for the linear mechanism and the triangular mechanism is enough information to permit calculation of each concentration as a function of time from the instant of initiation of the reaction until equilibrium is reached. We shall now develop explicit equations that can be used to calculate, for particular sets of rate constants, the time–concentration graphs for the linear mechanism.

Mechanism (5.2)–(5.3) yields the relaxation spectrum described by Equations (5.71), (5.72), and (5.73). We shall restrict our attention to experiments in which only A is present at the start of the reaction. Initial conditions are thus

$$(A)_0 = a_0 + a_1 + a_2 \tag{5.100}$$

$$(B)_0 = 0 = b_0 + b_1 + b_2 \tag{5.101}$$

$$(C)_0 = 0 = c_0 + c_1 + c_2 \tag{5.102}$$

where the subscript zero on the concentrations denotes $t = 0$. These three equations were obtained by setting t equal to zero in Equations (5.71)–(5.73), taking into account the fact that neither B nor C is present at the beginning of the reaction.

The equilibrium values of the three concentrations can be obtained from Equations (5.71)–(5.73) by finding the limit when t increases without upper bound. The results are

$$(\overline{A}) = a_0 \tag{5.103}$$

$$(\overline{B}) = b_0 \tag{5.104}$$

$$(\overline{C}) = c_0 \tag{5.105}$$

Conservation of material in the reaction vessel requires that

$$(A)_0 = (\overline{A}) + (\overline{B}) + (\overline{C}) \tag{5.106}$$

Substitution of Equations (5.103)–(5.105) into Equation (5.106) gives

$$(A)_0 = a_0 + b_0 + c_0 \tag{5.107}$$

Equations (5.63), (5.64), (5.103), (5.104), and (5.105) combine to yield

$$k_1 a_0 = k_{-1} b_0 \tag{5.108}$$

$$k_{-2} c_0 = k_2 b_0 \tag{5.109}$$

There are three fruitful ways of combining Equations (5.107), (5.108), and (5.109):

$$(A)_0 = a_0 \left(1 + \frac{k_1}{k_{-1}} + \frac{k_1 k_2}{k_{-1} k_{-2}} \right) \tag{5.110}$$

$$(A)_0 = b_0 \left(\frac{k_{-1}}{k_1} + 1 + \frac{k_2}{k_{-2}} \right) \tag{5.111}$$

$$(A)_0 = c_0 \left(\frac{k_{-1} k_{-2}}{k_1 k_2} + \frac{k_{-2}}{k_2} + 1 \right) \tag{5.112}$$

These equations permit numerical calculation of the three coefficients a_0, b_0, and c_0 from the initial concentration and the microscopic rate constants. We now proceed to find ways to calculate the remaining six coefficients.

After it has been found that there are three different values of m, Equation (5.12) can be evaluated for each m. In particular, for the two nonzero m's we can write

$$a_1[k_1 - m_1] = b_1 k_{-1} \tag{5.113}$$

$$a_2[k_1 - m_2] = b_2 k_{-1} \tag{5.114}$$

Combination of Equations (5.113), (5.114), (5.100), (5.101), (5.108), and (5.110) produces

$$a_1 = \frac{(A)_0 k_1}{m_2 - m_1}\left(1 - \frac{m_2[k_2 + k_{-2}]}{k_1 k_2 + k_{-1}k_{-2} + k_1 k_{-2}}\right) \tag{5.115}$$

$$a_2 = \frac{(A)_0 k_1}{m_2 - m_1}\left(1 - \frac{m_1[k_2 + k_{-2}]}{k_1 k_2 + k_{-1}k_{-2} + k_1 k_{-2}}\right) \tag{5.116}$$

Equations (5.113) and (5.114) can be used to transform (5.115) and (5.116) into

$$b_1 = \frac{(A)_0 k_1}{k_{-1}}\frac{k_1 - m_1}{m_1 - m_2}\left(1 - \frac{m_2[k_2 + k_{-2}]}{k_1 k_2 + k_{-1}k_{-2} + k_1 k_{-2}}\right) \tag{5.117}$$

$$b_2 = \frac{(A)_0 k_1}{k_{-1}}\frac{k_1 - m_2}{m_2 - m_1}\left(1 - \frac{m_1[k_2 + k_{-2}]}{k_1 k_2 + k_{-1}k_{-2} + k_1 k_{-2}}\right) \tag{5.118}$$

Evaluation of Equation (5.14) for $m = m_1$ and for $m = m_2$ yields

$$b_1 k_2 = c_1[k_{-2} - m_1] \tag{5.119}$$

$$b_2 k_2 = c_2[k_{-2} - m_2] \tag{5.120}$$

Then Equations (5.119) and (5.120) can be substituted into (5.117) and (5.118) to give

$$c_1 = \frac{(A)_0 k_1 k_2}{k_{-1}}\frac{k_1 - m_1}{k_{-2} - m_1}\frac{1}{m_1 - m_2}\left(1 - \frac{m_2[k_2 + k_{-2}]}{k_1 k_2 + k_{-1}k_{-2} + k_1 k_{-2}}\right) \tag{5.121}$$

$$c_2 = \frac{(A)_0 k_1 k_2}{k_{-1}}\frac{k_1 - m_2}{k_{-2} - m_2}\frac{1}{m_2 - m_1}\left(1 - \frac{m_1[k_2 + k_{-2}]}{k_1 k_2 + k_{-1}k_{-2} + k_1 k_{-2}}\right) \tag{5.122}$$

Equations (5.110), (5.111), (5.112), (5.115), (5.116), (5.117), (5.118), (5.121), and (5.122) can be written in more compact forms by introducing Equations (5.29) and (5.30). The resulting equations are

$$a_0 = (A)_0 \frac{k_{-1}k_{-2}}{m_1 m_2} \tag{5.123}$$

$$b_0 = (A)_0 \frac{k_1 k_{-2}}{m_1 m_2} \tag{5.124}$$

$$c_0 = (A)_0 \frac{k_1 k_2}{m_1 m_2} \tag{5.125}$$

$$a_1 = (A)_0 k_1 \frac{m_1 - k_2 - k_{-2}}{m_1[m_1 - m_2]} \tag{5.126}$$

$$a_2 = (A)_0 k_1 \frac{m_2 - k_2 - k_{-2}}{m_2[m_2 - m_1]} \tag{5.127}$$

$$b_1 = (A)_0 k_1 \frac{k_{-2} - m_1}{m_1[m_1 - m_2]} \tag{5.128}$$

$$b_2 = (A)_0 k_1 \frac{k_{-2} - m_2}{m_2[m_2 - m_1]} \tag{5.129}$$

$$c_1 = (A)_0 \frac{k_1 k_2}{m_1[m_1 - m_2]} \tag{5.130}$$

$$c_2 = (A)_0 \frac{k_1 k_2}{m_2[m_2 - m_1]} \tag{5.131}$$

The concentration of reactant A is initially $(A)_0$. This concentration decreases during the reaction, asymptotically approaching $(\overline{A})$ at very large values of t. The explicit equation for the time course of (A) is found by substituting Equations (5.123), (5.126), and (5.127) into Equation (5.71). The result is

$$(A) = (A)_0 \left(\frac{k_{-1} k_{-2}}{m_1 m_2} + \frac{k_1[m_1 - k_2 - k_{-2}]e^{-m_1 t}}{m_1[m_1 - m_2]} + \frac{k_1[m_2 - k_2 - k_{-2}]e^{-m_2 t}}{m_2[m_2 - m_1]} \right) \tag{5.132}$$

The concentration of product C increases continuously as the reaction progresses. With C there is in sequence an initial induction period, a period of accelerated rate of reaction, an inflection point, and finally an asymptotic approach to the equilibrium value $(\overline{C})$. The time course of the concentration (C) is found by substituting values of the coefficients given by Equations (5.125), (5.130), and (5.131) into Equation (5.73), producing

$$(C) = (A)_0 k_1 k_2 \left(\frac{1}{m_1 m_2} + \frac{e^{-m_1 t}}{m_1[m_1 - m_2]} + \frac{e^{-m_2 t}}{m_2[m_2 - m_1]} \right) \tag{5.133}$$

Several features of this function can be seen with greater clarity by examining the slope of the (C) versus t plot. This slope is found by differentiating Equation (5.133) with respect to t, giving

$$\frac{d(C)}{dt} = \frac{(A)_0 k_1 k_2}{m_1 - m_2} (-e^{-m_1 t} + e^{-m_2 t}) \qquad (5.134)$$

Since $m_1 > m_2$, this derivative can never be negative. The difference $[m_1 - m_2]$ is positive, and the exponential $e^{-m_1 t}$ can never be larger than the exponential $e^{-m_2 t}$. We can check to see if the function passes through a maximum or a minimum, at any point from the very beginning of the reaction through the approach to equilibrium, by finding the conditions, if any, under which the derivative $d(C)/dt$ is equal to zero. When the derivative is zero, the slope of the curve is zero, and the curve is at either a maximum or a minimum value. The requirement that $d(C)/dt = 0$ is equivalent to

$$e^{-m_1 t} = e^{-m_2 t} \qquad (5.135)$$

This equality can be valid only at the initial instant when $t = 0$. This function begins with a zero slope, with the plot of (C) versus t parallel to the time axis. At all subsequent values of t, the slope is positive, and thus the value of (C) continues to rise throughout the reaction.

The intermediate B is absent at the start of the reaction. The value of (B), initially zero, may rise continuously toward its limiting value $(\overline{B})$, or it may pass through a maximum. The concentration of the intermediate B varies with time according to the equation found by substituting Equations (5.124), (5.128), and (5.129) into Equation (5.72):

$$(B) = (A)_0 k_1 \left(\frac{k_{-2}}{m_1 m_2} + \frac{k_{-2} - m_1}{m_1 [m_1 - m_2]} e^{-m_1 t} + \frac{k_{-2} - m_2}{m_2 [m_2 - m_1]} e^{-m_2 t} \right)$$

$$(5.136)$$

Again we will examine the slope of the concentration versus time curve throughout the course of the reaction. Differentiation of Equation (5.136) with respect to t yields

$$\frac{d(B)}{dt} = \frac{(A)_0 k_1}{m_1 - m_2} ([m_1 - k_{-2}]e^{-m_1 t} - [m_2 - k_{-2}]e^{-m_2 t}) \qquad (5.137)$$

This derivative is equal to zero, and therefore (B) is at a maximum or a minimum value, when

$$[m_1 - k_{-2}]e^{-m_1 t} = [m_2 - k_{-2}]e^{-m_2 t} \tag{5.138}$$

Both exponentials are always positive. Thus Equation (5.138) can be satisfied only if the factors in parentheses have the same sign. The requirement for a maximum or a minimum is that k_{-2} be greater than both m's, or that k_{-2} be less than both m's. (See Problem 5.2, which deals with the possibility that any microscopic rate constant can be larger than both m's.) Otherwise the function will be an increasing function, beginning with (B) = 0 at $t = 0$, and continuing as (B) approaches the limiting value $(\overline{B})$ at very large values of t. The more interesting case occurs when k_{-2} is less than m_2. Then the concentration of the intermediate rises during the first phase of the reaction, reaches a maximum value, and then decreases asymptotically toward $(\overline{B})$. Because of the fact that even in this simple mechanism, it is possible for the concentration of a reacting species to pass past its equilibrium value, and in fact to perform a primitive sort of oscillation about its equilibrium value (a critically damped oscillation), it is important to be careful in making statements about the impossibility of overshooting equilibrium. For while it is generally considered impossible for an isolated system to overshoot equilibrium, it is nevertheless possible to find that individual concentrations oscillate about their equilibrium values as the system as a whole relaxes toward, but not past, equilibrium. The more complicated the mechanism, the more possibilities there are for complicated oscillations of individual concentrations about the equilibrium point.

Problems

5.1. Consider Mechanism (5.2)–(5.3) with k_1 and k_2 individually much greater than k_{-1} and k_{-2}. Write (B) as a function of time, with coefficients of the exponentials evaluated in terms of the initial concentration and the k's, for the two cases $k_1 \gg k_2$ and $k_2 \gg k_1$. Consider that at initial time, (A) = (A)$_0$, (B) = 0, (C) = 0.

5.2. For Mechanism (5.2)–(5.3), can the value of k_{-2} be greater than the value of m_1? Begin by assuming numerical values for the k's, and calculate the corresponding m's.

5.3. Equating the derivative $d(C)/dt$ to zero gives an equation valid for each point in a plot of (C) versus t at which the function passes through a maximum or a minimum value. When the second derivative, the time derivative of $d(C)/dt$, is set equal to zero, the resulting equation is valid for each point at which the (C) versus t curve shows an inflection. Find the conditions, if any, which result in an inflection in plots of (B) versus t, and in plots of (C) versus t, for Mechanism (5.2)–(5.3).

5.4. Find the relationships among the three nonzero m's and the six k's for the mechanism

$$A \underset{k_{-1}}{\overset{k_1}{\rightleftharpoons}} B$$

$$B \underset{k_{-2}}{\overset{k_2}{\rightleftharpoons}} C$$

$$C \underset{k_{-3}}{\overset{k_3}{\rightleftharpoons}} D$$

5.5. Find the relationships among the three nonzero m's and the six k's for the mechanism

$$A \underset{k_{-1}}{\overset{k_1}{\rightleftharpoons}} B$$

$$A \underset{k_{-2}}{\overset{k_2}{\rightleftharpoons}} C$$

$$A \underset{k_{-3}}{\overset{k_3}{\rightleftharpoons}} D$$

5.6. When optically-active $[Co\ en_2Cl_2]Cl$ is dissolved in water, the resulting solution is optically active. Chemical reactions occur, and the optical rotation α of the solution, measured with a polarimeter, changes with time. In one such experiment, the following student data were obtained:

$t\ (min)$	$\alpha\ (deg)$	$t\ (min)$	$\alpha\ (deg)$
0	0.201	30	0.129
3	0.188	36	0.123
6	0.180	45	0.108
9	0.175	60	0.095
12	0.168	90	0.076
15	0.162	120	0.065
18	0.156	180	0.055
21	0.147	300	0.032
27	0.135	1440	0.000

Assuming that these data result from the superposition of two first-order relaxations, determine the values of two macroscopic first-order rate constants.

5.7. A half-life of 10^{-9} second is certainly short, perhaps too short for the unassisted imagination to contemplate. The imagination may get an assist via a comparison with how long 10^9 seconds is. Convert the following times into seconds, retaining only one or two significant figures: 1 hour; 1 day; 1 year; 1 century.

5.8. Show that the time-lag method can be used for evaluating either m in Equation (5.76) if the two m's have sufficiently different numerical values.

BIBLIOGRAPHICAL NOTE

A concise introduction to chemical relaxation is given by D. N. Hague, *Relaxation Methods for Studying Very Rapid Reactions in Solution*, London: The Royal Institute of Chemistry, 1964. This 16-page pamphlet is publication No. 4 of the 1964 Royal Institute Lecture Series.

The general solution for the reaction sequence

$$M_1 \rightleftharpoons M_2 \rightleftharpoons \cdots \rightleftharpoons M_n \rightleftharpoons M_{n+1}$$

was developed by A. Rakowski [*Z. Physik. Chem.* (Leipzig), **57,** 321 (1907)]. A detailed analysis of the solutions of rate equations for Mechanism (5.2)–(5.3) is given by T. M. Lowry and W. T. John [*J. Chem. Soc.*, **97,** 2634 (1910)].

Explicit equations for the time course of the concentrations in Mechanism (5.84)–(5.86) have been presented for several limiting cases in which two microscopic rate constants are very small with respect to the remaining four k's, and also for some cases in which relationships are imposed among certain k's, by R. A. Alberty and W. G. Miller [*J. Chem. Phys.*, **26,** 1231 (1957)].

Two very complete books dealing with the mathematics of coupled simultaneous reactions, and the resulting chemical relaxation processes, are G. H. Czerlinski, *Chemical Relaxation*, New York: Marcel Dekker, Inc., 1966; and N. M. Rodiguin and E. N. Rodiguina, *Consecutive Chemical Reactions*, New York: Van Nostrand Reinhold Company, 1964, English edition edited by R. F. Schneider.

The differential rate equations for a set of coupled first-order chemical reactions are identical in form with the equations describing vibration of a polyatomic molecule. F. A. Matsen and J. L. Franklin [*J. Amer. Chem. Soc.*, **72,** 3337 (1950)] carry out the mathematical analysis of rate equations in the same manner as is customary in dealing with normal modes of vibration. They point out the analogies between normal coordinates for vibration of a linear triatomic molecule, and quantities, which they term eigenconcentrations, arising in their treatment of Reaction Mechanism (5.2)–(5.3).

A detailed analysis of cyclic Reaction Mechanism (5.84)–(5.86) from a geometric point of view, taking advantage of the power of matrix algebra, forms a substantial portion of the 190-page chapter, "The Structure and Analysis of Complex Reaction Systems," by J. Wei and C. D. Prater, in *Advances in Catalysis*, vol. 13, New York: Academic Press, Inc., 1962, p. 203. They describe the composition of the reacting system at any particular time as a vector. This vector, for a three-component system, is constrained within a reaction triangle. As reaction proceeds, the end of the vector traces out a curve which Wei and Prater call the reaction path. They emphasize experimental conditions under which these reaction paths are straight lines. This elegant method of Wei and Prater is nicely surveyed in M. Boudart, *Kinetics of Chemical Processes*, Englewood Cliffs, N.J.: Prentice-Hall, Inc., 1968, pp. 215–237, where there is a section on the determination of straight-line reaction paths.

Solution of the set of simultaneous rate equations is an application of mathematical eigenvalue theory. Eigenvalue theory also has important chemical applications in the theory of molecular vibrations and in the theory of molecular orbitals. An extensive treatment of eigenvalue theory is given by C. Lanczos, *Applied Analysis*, Englewood Cliffs, N.J.: Prentice-Hall, Inc., 1956. Chapter II, "Matrices and Eigenvalue Problems," is particularly helpful.

6

Reversibility

Each reaction mechanism in this book is written as a set of elementary chemical reactions, and every one of those elementary reactions is written with a double arrow to signify reversibility. The requirement that each elementary reaction be reversible is justified in Chapter 1 by an appeal to the principle of microscopic reversibility. What is this principle? And why should a chemist believe in microscopic reversibility?

Suppose that there were a gas-phase reaction in which the compound X is converted to Y and Z by an irreversible reaction step

$$X \to Y + Z \qquad (6.1)$$

This reaction results in an increase in the number of molecules. Suppose that a solid catalyst, D, is found which reacts with Y and Z to produce X, D undergoing no net change during the reaction. Catalyst D has no effect on Reaction (6.1). Let the reaction proceed in a piston-and-cylinder arrangement such as the one pictured in Figure 6.1.

This gas-tight cylinder thus contains gaseous molecules X, Y, and Z throughout the apparatus, and solid compound D in the little box. These compounds can interconvert according to the mechanism

$$X \xrightarrow{k_1} Y + Z \qquad (6.2)$$

$$D + X \xleftarrow[k_{-2}]{} Y + Z + D \qquad (6.3)$$

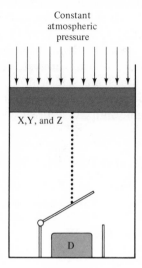

FIGURE 6.1. A perpetual-motion machine.

When the lid of the little box is closed, X reacts via (6.2) to give Y and Z; the number of molecules in the cylinder increases. An increase in the number of molecules at constant pressure requires an increase in volume, and thus the piston rises, opening the lid of the little box. When the lid is opened, the catalyst is exposed to the reacting gases. Reaction (6.3) begins, reforming X and reducing the number of molecules in the cylinder. The piston falls, closing the lid of the little box, and one cycle has been completed. If D remains solid and stays in the box, the whole system is unchanged after completion of the cycle. The system is prepared to continue spontaneously through the cycle again. The piston could be connected to another machine so that useful work could be done. This is perpetual motion with a capacity for doing useful work forever. There is an unlimited output of work without any input of energy.

If the perpetual motion machine were to work, it would perform its work in violation of the law of conservation of energy. Something is wrong with assumptions that led to the prediction of perpetual motion.

There is, in fact, something fundamentally wrong with a kinetic reaction mechanism which has a one-way microscopic reaction step that is absolutely irreversible.

The principle of microscopic reversibility, sometimes called the principle of detailed balancing, requires rejection of both elementary reactions (6.2) and (6.3) simply because they are written and interpreted as strictly irreversible. This principle requires that, when equilibrium is established in a chemical system, it be established in a *pairwise sense* between each set of products and reactants which can interconvert in a single microscopic reaction step. When applied to Mechanism (6.2)–(6.3), the principle of microscopic reversibility requires double arrows for both reactions. At equilibrium, in the presence of the catalyst, both reactions must be at equilibrium. These conditions are sufficient to assure that the catalyst, now a catalyst for both forward and reverse elementary processes, cannot affect the position of equilibrium. As a direct consequence, the entire perpetual motion scheme fails.

Microscopic Reversibility in a Triangular Mechanism

Several important consequences of the principle of microscopic reversibility can be readily illustrated by considering the triangular isomerization mechanism in which compounds A, B, and C are interconverted via unimolecular elementary reactions. Thus

$$
\begin{array}{c}
\text{A} \\[2em]
\substack{k_3 \\ k_{-3}} \qquad \substack{k_1 \\ k_{-1}} \\[2em]
\text{C} \;\; \underset{k_{-2}}{\overset{k_2}{\rightleftarrows}} \;\; \text{B}
\end{array}
\qquad (6.4)
$$

Let us examine this system in a state of chemical equilibrium. At equilibrium the six microscopic reactions continue to proceed, but each of the macroscopic chemical concentrations has a value that remains unchanged in time. The amount of A being produced is balanced by the amount of A being consumed. This is dynamic equilibrium, a state in which the macroscopic properties of the reacting system remain unchanged as time marches on, but in which the molecules themselves are involved in continuing chemical transformations. There is a great deal of activity on the molecular level.

There is another kinetic reaction mechanism that could give rise to the same macroscopic state of chemical equilibrium. Consider the triangular mechanism in which each molecular transformation is considered to be strictly irreversible:

$$
\begin{array}{c}
\text{A} \\[2em]
k_6 \qquad k_4 \\[2em]
\text{C} \;\; \overset{k_5}{\longleftarrow} \;\; \text{B}
\end{array}
\qquad (6.5)
$$

An appropriate choice of rate constants could yield precisely the same equilibrium chemical concentrations for the two different reaction mechanisms. Yet Mechanism (6.5) has some disturbing aspects. The molecules are on a reaction merry-go-round, traveling on a one-way reaction pathway. If in some way it were possible to inhibit the reaction

$$A \xrightarrow{k_4} B$$

then eventually all molecules would be A. Or if some catalyst were introduced which increased the numerical value of k_4, then more B would accumulate than before the rate constant was changed; the relative values of the macroscopic equilibrium concentrations of A, B, and C would be altered.

The general problem with Mechanism (6.5) is that it predicts a state of equilibrium in which the chemical concentrations are critically dependent on the details of the mechanism whereby equilibrium was achieved. The presence of a catalyst (which really means the existence of an alternative pathway for the reaction) would change equilibrium properties of the system. But the equilibrium properties must be determined by the value of the equilibrium constant, not by the particular kinetic mechanism by which equilibrium is achieved.

A microscopic equilibrium constant must exist for each microscopic reaction step. Thus the relative equilibrium concentrations of directly interconvertible chemical species can be considered as independent of all other reactions. The existence of alternative pathways for an overall reaction will most likely affect the rate of reaction, but it cannot affect the relative equilibrium concentrations.

Application of the principle of microscopic reversibility to the detailed balancing of the three reaction steps in Mechanism (6.4) results in the conclusion that not all six rate constants can be independent. The same equilibrium concentrations of A and B must result from either of the two alternative reaction pathways; the direct pathway characterized by rate constants k_1 and k_{-1} must yield the same equilibrium concentrations as the pathway, through the intermediate species C, which involves the four rate constants k_2, k_{-2}, k_3, and k_{-3}.

Rate Constants, Equilibrium Constants, and Microscopic Reversibility. We again look at the triangular mechanism (6.4) with the complete set of six rate constants, and try to find the relationships among these rate constants and the equilibrium constants required by the principle of microscopic reversibility. This discussion follows a treatment[1] given by the theo-

---

[1] L. Onsager, *Phys. Rev.*, [2] **37**, 410 (1931).

retical physical chemist, Lars Onsager, who was awarded the Nobel Prize in chemistry in 1968.

There are three differential rate equations associated with this mechanism:

$$\frac{d(A)}{dt} = -[k_1 + k_{-3}](A) + k_{-1}(B) + k_3(C) \tag{6.6}$$

$$\frac{d(B)}{dt} = -[k_{-1} + k_2](B) + k_{-2}(C) + k_1(A) \tag{6.7}$$

$$\frac{d(C)}{dt} = -[k_{-2} + k_3](C) + k_2(B) + k_{-3}(A) \tag{6.8}$$

At equilibrium, the derivatives $d(A)/dt$, $d(B)/dt$, and $d(C)/dt$ must individually be equal to zero. Equations (6.6), (6.7), and (6.8) evaluated at equilibrium yield

$$(\overline{A}) = \frac{k_{-1}(\overline{B}) + k_3(\overline{C})}{k_1 + k_{-3}} \tag{6.9}$$

$$(\overline{B}) = \frac{k_{-2}(\overline{C}) + k_1(\overline{A})}{k_{-1} + k_2} \tag{6.10}$$

$$(\overline{C}) = \frac{k_2(\overline{B}) + k_{-3}(\overline{A})}{k_{-2} + k_3} \tag{6.11}$$

Equations (6.11) and (6.9) can be combined to eliminate $(\overline{C})$ and to obtain an expression for the usual equilibrium-constant quotient $(\overline{B})/(\overline{A})$:

$$\frac{(\overline{B})}{(\overline{A})} = \frac{k_1 k_{-2} + k_{-3} k_{-2} + k_1 k_3}{k_3 k_2 + k_3 k_{-1} + k_{-1} k_{-2}} \tag{6.12}$$

From Equations (6.10) and (6.9) comes

$$\frac{(\overline{C})}{(\overline{B})} = \frac{k_1 k_2 + k_{-3} k_{-1} + k_{-3} k_2}{k_{-2} k_1 + k_{-2} k_{-3} + k_1 k_3} \tag{6.13}$$

Equations (6.10) and (6.11) together give

$$\frac{(\overline{A})}{(\overline{C})} = \frac{k_3 k_2 + k_3 k_{-1} + k_{-1} k_{-2}}{k_1 k_2 + k_{-3} k_{-1} + k_{-3} k_2} \tag{6.14}$$

Equations (6.12), (6.13), and (6.14) are not independent. Any two of the equations can be used to obtain the third.

These three equations constitute a strange set of criteria for chemical equilibrium. They do not have the form of the familiar equilibrium-constant equations. In fact, these equations could be satisfied by a nonequilibrium system in which the steady-state, time-invariant concentrations were maintained by a clockwise irreversible mechanism involving rate constants k_1, k_2, and k_3, together with an independent counterclockwise irreversible mechanism involving rate constants k_{-1}, k_{-2}, and k_{-3}.

"Here, however, the chemists are accustomed to impose a very interesting additional restriction, namely: when the equilibrium is reached, each individual reaction must balance itself. They require that the transition $A \rightarrow B$ must take place just as frequently as the reverse transition $B \rightarrow A$ etc." (L. Onsager; see footnote 1). This additional restriction, a consequence of the principle of microscopic reversibility, rejects the possibility of equilibrium being maintained by independent clockwise and counterclockwise processes. This detailed balancing is achieved by considering each of the reversible elementary reactions as independently able to bring about chemical equilibrium between the reactant and product in that reaction.

The derivative $d(A)/dt$ must be zero at equilibrium. When chemical equilibrium has been achieved, equilibrium must exist in a pairwise sense between A and B, and the correct microscopic equilibrium constant will be obtained by looking just at the single elementary reaction

$$A \underset{k_{-1}}{\overset{k_1}{\rightleftharpoons}} B \qquad (6.15)$$

At equilibrium, the differential rate equation in terms of the time dependence of (A) must be

$$\frac{d(A)}{dt} = 0 = -k_1(\overline{A}) + k_{-1}(\overline{B}) \qquad (6.16)$$

which looks more familiar in the form

$$\frac{k_1}{k_{-1}} = \frac{(\overline{B})}{(\overline{A})} \qquad (6.17)$$

Equation (6.17) is the definition of the microscopic equilibrium constant for Reaction (6.15), written in terms of concentrations. Thus we have

$$K_1' \equiv \frac{k_1}{k_{-1}} = \frac{(\overline{B})}{(\overline{A})} \qquad (6.18)$$

By the same reasoning

$$K_2' \equiv \frac{k_2}{k_{-2}} = \frac{(\overline{C})}{(\overline{B})} \tag{6.19}$$

$$K_3' \equiv \frac{k_3}{k_{-3}} = \frac{(\overline{A})}{(\overline{C})} \tag{6.20}$$

Equations (6.18), (6.19), and (6.20) can be combined to give

$$k_1 k_2 k_3 = k_{-1} k_{-2} k_{-3} \tag{6.21}$$

$$K_1' K_2' K_3' = 1 \tag{6.22}$$

Equations (6.21) and (6.22) show that of the three equilibrium constants, only two are independent; and that of the six rate constants, only five are independent. If the numerical values of two equilibrium constants are known, the value of the third can be calculated. If five rate constants are known, the sixth can be calculated.

The requirement of microscopic reversibility can be imposed on Equations (6.12), (6.13), and (6.14) in the form of restriction (6.21). Substitution of (6.21) into Equation (6.12) results in

$$\frac{(\overline{B})}{(\overline{A})} = \frac{k_1 k_{-2} + k_1 k_3 + k_{-2}[k_1 k_2 k_3 / k_{-1} k_{-2}]}{k_3 k_2 + k_3 k_{-1} + k_{-1} k_{-2}}$$

$$= \frac{k_1 k_{-2} k_{-1} + k_1 k_3 k_{-1} + k_1 k_2 k_3}{k_3 k_2 k_{-1} + k_3 k_{-1}{}^2 + k_{-1}{}^2 k_{-2}}$$

$$= \frac{k_1}{k_{-1}} \left(\frac{k_{-2} k_{-1} + k_{-1} k_3 + k_2 k_3}{k_{-2} k_{-1} + k_{-1} k_3 + k_2 k_3} \right) = \frac{k_1}{k_{-1}} \equiv K_1' \tag{6.23}$$

Equations (6.19) and (6.20) can be obtained in the same way.

Catalysis and Microscopic Reversibility

It is commonly said that a catalyst affects the rate of a chemical reaction, but that it cannot affect the final equilibrium state. Inasmuch as a catalyst serves only to provide an alternative reaction pathway from reactants to products, it is forbidden by the principle of microscopic reversibility to have any influence on the composition of the equilibrium state.

Consider the interconversion of M and N according to a reaction mechanism that provides two alternative pathways:

$$M \underset{k_{-1}}{\overset{k_1}{\rightleftharpoons}} N \qquad (6.24)$$

$$L + M \underset{k_{-2}}{\overset{k_2}{\rightleftharpoons}} N + L \qquad (6.25)$$

The compound L serves as a catalyst for the reaction by providing (6.25) as an alternative pathway which may, depending on the values of the rate constants, permit a much faster reaction rate than (6.24) alone.

The differential rate equation for the concentration (M) is

$$\frac{d(M)}{dt} = -k_1(M) - k_2(L)(M) + k_{-1}(N) + k_{-2}(L)(N) \qquad (6.26)$$

Evaluated at equilibrium, with $d(M)/dt = 0$, Equation (6.26) is

$$(\overline{M})[k_1 + k_2(\overline{L})] = (\overline{N})[k_{-1} + k_{-2}(\overline{L})] \qquad (6.27)$$

which can be written as

$$\frac{(\overline{N})}{(\overline{M})} = \frac{k_1 + k_2(\overline{L})}{k_{-1} + k_{-2}(\overline{L})} \qquad (6.28)$$

This is not the expected form of the equilibrium constant, because the conventional equilibrium constants for both (6.24) and (6.25) are independent of the concentration of L. Once again, the additional requirement of the principle of microscopic reversibility is needed. It is necessary that, at equilibrium, the elementary reactions (6.24) and (6.25) individually be in equilibrium, so that

$$k_1(\overline{M}) = k_{-1}(\overline{N}) \qquad (6.29)$$

$$k_2(\overline{M})(\overline{L}) = k_{-2}(\overline{N})(\overline{L}) \qquad (6.30)$$

Division of Equation (6.29) by Equation (6.30) results in the following requirement of interdependence of the rate constants:

$$k_1 k_{-2} = k_2 k_{-1} \qquad (6.31)$$

Substitution of Equation (6.31) into Equation (6.28) gives

$$\frac{(\overline{N})}{(\overline{M})} = \frac{k_1 + (\overline{L})k_1 k_{-2}/k_{-1}}{k_{-1} + k_{-2}(\overline{L})} = \frac{k_1 k_{-1} + k_1 k_{-2}(\overline{L})}{k_{-1}^2 + k_{-1} k_{-2}(\overline{L})}$$

$$= \frac{k_1}{k_{-1}}\left(\frac{k_{-1} + k_{-2}(\overline{L})}{k_{-1} + k_{-2}(\overline{L})}\right) = \frac{k_1}{k_{-1}} \equiv K_1' \tag{6.32}$$

The effect of the principle of microscopic reversibility is seen, not just at equilibrium, but also during the entire reaction, through the requirements imposed on relative values of the several microscopic rate constants of the reaction mechanism.

Problem

6.1. Show that the restriction (6.31) imposed by the principle of microscopic reversibility is sufficient to convert Equation (6.28) to

$$\frac{(\overline{N})}{(\overline{M})} = \frac{k_2}{k_{-2}}$$

BIBLIOGRAPHICAL NOTE

The fact that equations such as the set (6.12)–(6.14) do not constitute a satisfactory criterion for chemical equilibrium was noticed early in the history of chemical kinetics. R. Wegscheider published a detailed paper on the subject [*Z. Physik. Chem.* (Leipzig), **39**, 257 (1902)], and the problem became known as Wegscheider's paradox. This paradox was resolved in essentially the manner presented in this chapter by G. N. Lewis [*Proc. Nat. Acad. Sci. U.S.*, **11**, 179 (1925)]. Lewis argued that the cyclic equilibrium of a mechanism such as (6.5) must be impossible, and was

led to propose a law which in its general form is not deducible from thermodynamics, but proves to be compatible with the laws of thermodynamics in all cases where a comparison is possible. It may be called the law of *entire equilibrium,* and may be stated as follows. *Corresponding to every individual process there is a reverse process, and in a state of equilibrium the average rate of every process is equal to the average rate of its reverse process...* The law of entire equilibrium might have been called the law of reversibility to the last detail.

7

The Bimolecular Reaction
of Piperidine and
2,4-Dinitrochlorobenzene:

A Case Study

The compound 2,4-dinitrochlorobenzene,

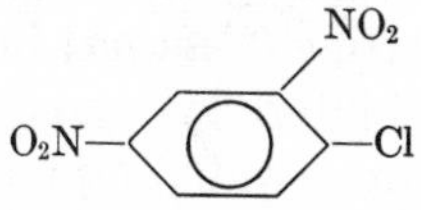

reacts in ethanol with the compound piperidine,

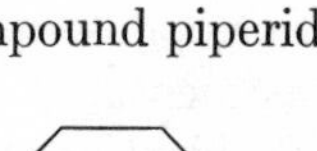

to form, with loss of a proton and a chloride ion, the compound 2,4-dinitro-phenylpiperidine,

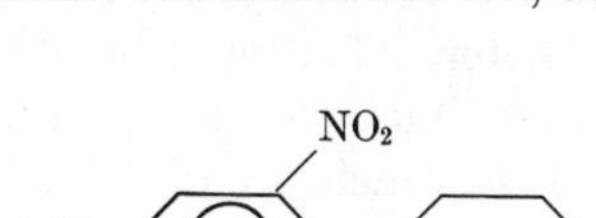

It is thought that the ions H^+ and Cl^- are transferred to another piperidine

molecule to yield the piperidine hydrochloride,

$$\langle\ \rangle NH_2^+Cl^-$$

P-HCl

Reactant 2,4-dinitrochlorobenzene is a white solid that dissolves readily in ethanol to give a colorless solution. Piperidine is a pungent, colorless liquid which also yields a colorless solution in ethanol. When ethanolic solutions of the two reactants are mixed, a slow color change occurs which can be observed visually. The solution becomes yellow, the color being quite pronounced even when reactant concentrations are of the order of 10^{-3} molar or less.

When an experiment is designed so that the initial concentration of piperidine is substantially greater than the initial concentration of 2,4-dinitrochlorobenzene, the appearance of yellow color is a first-order relaxation process characterized by a macroscopic first-order rate constant which we shall call m_1. The numerical value of m_1 depends on the temperature and on the initial concentration of piperidine.

Preliminary Experiment. A solution was prepared by dissolving 80 mg of piperidine in absolute ethanol, diluting to 100 ml in a volumetric flask. A second solution was prepared by dissolving 20 mg of 2,4-dinitrochlorobenzene in absolute ethanol, diluting to 100 ml in a volumetric flask. The reaction was initiated by pipetting **(Caution! Do not pipette these solutions by mouth!** Use some type of bulb or syringe pipetting aid, or use a length of rubber tubing attached to a water aspirator, to draw the solutions into the pipettes) 10.00 ml of piperidine solution and 5.00 ml of 2,4-dinitrochlorobenzene solution into a small beaker. The mixture was stirred and then transferred into a spectrophotometer cell with an optical path length of 1 cm. Absorbance was measured continuously at a wavelength of 360 mμ for 2 hours, using a Perkin-Elmer Model 350 recording spectrophotometer. The spectrophotometer plot is presented in Figure 7.1.

These absorbance values are said to be unknown within an arbitrary additive constant. For convenience, the recorder pen was set near the position of zero absorbance at the beginning of the reaction by turning a control knob on the spectrophotometer. This knob has the effect of shifting the entire absorbance scale back and forth across the paper. The recording paper moved at a rate of 0.0865 inch per minute. In spite of the fact that the zero point of the absorbance scale is unknown, and in spite of the fact that the time scale is in inconvenient units, the rate curve can be easily analyzed by the Guggenheim method if the curve is, in fact, a first-order relaxation curve. The unknown additive constant disappears when absorbance differences are taken. The plot can be made in terms of time ex-

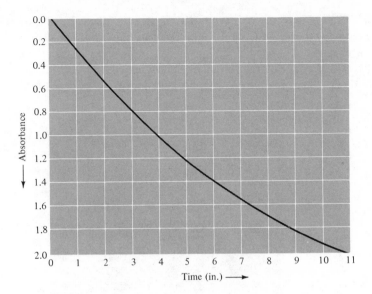

FIGURE 7.1.   Plot of absorbance versus time for the preliminary experiment.

pressed as "chart divisions" or inches. The conversion factor between length and time, between inches and minutes, is needed later. A sample calculation is presented in Table 7.1, and the resulting Guggenheim plot appears in Figure 7.2.

The wavelength and the cell path length have no effect on the value of the rate constant, but both should be chosen so as to take full advantage of the accuracy and range of the spectrophotometer. The rate plot should cover almost the full width of the recorder chart. The absorbance change is smaller at longer wavelengths.

Designing a Series of Kinetic Runs.   Using the preliminary information, experiments can be designed to permit evaluation of the rate constants for some possible mechanisms. It is emphasized throughout Chapter 8 that bimolecular elementary reactions yield concentration-dependent macroscopic rate constants. Therefore, information about the dependence of the value of the macroscopic first-order rate constant on the initial concentration of piperidine is needed. It is unlikely that small errors in the concentration of 2,4-dinitrochlorobenzene will have a noticeable effect on the value of the observed rate constant, but it is important to know the value of the initial concentration of piperidine with enough accuracy so that uncertainties in the values of macroscopic rate constants are the limiting source of error in the quantitative interpretation of the experiments. The value of the observed rate constant changes by almost a factor of 10 between 0 and 25°C, and so for accurate measurements some manner of thermostatting is needed. If longer cell path lengths are available, and if small absorbance differences can be determined with the available spectrophotometer, it is possible to decrease substantially the 2,4-dinitrochloro-

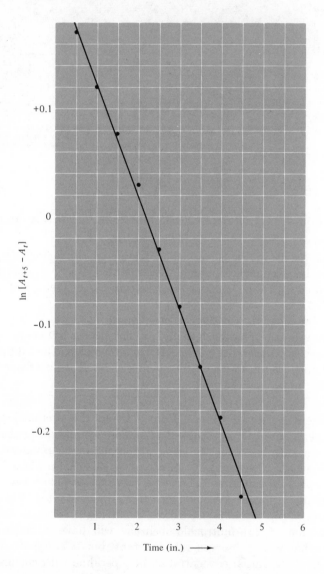

FIGURE 7.2. Determination of macroscopic rate constant using a Guggenheim plot.

TABLE 7.1.   Calculations for Constructing a Guggenheim Plot

t $(in.)$	A_t	A_{t+5}	$A_{t+5} - A_t$	$ln\{A_{t+5} - A_t\}$
0.5	0.11	1.30	1.19	0.174
1.0	0.26	1.39	1.13	0.122
1.5	0.40	1.48	1.08	0.077
2.0	0.53	1.56	1.03	0.030
2.5	0.66	1.63	0.97	−0.030
3.0	0.78	1.70	0.92	−0.083
3.5	0.90	1.77	0.87	−0.139
4.0	1.00	1.83	0.83	−0.186
4.5	1.11	1.88	0.77	−0.261
5.0	1.21			
5.5	1.30			
6.0	1.39			
6.5	1.48			
7.0	1.56			
7.5	1.63			
8.0	1.70			
8.5	1.77			
9.0	1.83			
9.5	1.88			

benzene concentrations. Such a reduction widens the range of accessible piperidine concentrations, and thus the range in which the concentration dependence of m_1 on $(P)_0$ can be determined. The value of $(P)_0$ must be kept large with respect to $(B)_0$ in the particular kinetic run. Otherwise, the only restrictions are practical considerations of the instrumental problem of accurate measurement of the observed absorbance change, of being able to mix reactants fast enough when the reaction is rapid, and of being able to monopolize the spectrophotometer for a long period of time when the reaction is very slow.

Proposals about Mechanism.   One proposed mechanism[1] for this reaction can be formulated as

$$B + P \underset{k_{-1}}{\overset{k_1}{\rightleftharpoons}} C\text{-HCl} \tag{7.1}$$

$$P + C\text{-HCl} \underset{k_{-2}}{\overset{k_2}{\rightleftharpoons}} C + P\text{-HCl} \tag{7.2}$$

[1] J. F. Bunnett and H. D. Crockford, *J. Chem. Educ.*, **33,** 552 (1956).

where the intermediate species C-HCl is 2,4-dinitrophenylpiperidine hydrochloride, an ion pair in which a chloride anion is closely associated with the cation

The mathematical methods of Chapter 8 can be used to show that this mechanism should yield first-order rates if suitable experimental conditions can be arranged so that (P) and (P-HCl) remain constant throughout the reaction. A large initial excess of piperidine assures us that $(P) \simeq (P)_0$ throughout the reaction. It is not altogether clear, however, that (P-HCl) is necessarily constant. Perhaps its effect on the differential equations is in algebraic terms that are insignificantly small. If not, it may be necessary to keep this concentration constant by the addition of P-HCl to the initial reaction mixture. How could this be done? Can enough data be obtained from these experiments to evaluate one or more rate constants of the mechanism? If some cannot be evaluated, is it possible to place upper or lower bounds on their values?

This reaction has significance within a larger framework of reaction mechanisms, and several related reactions are discussed by Bunnett, Garbisch, and Pruitt.[2] There are two aspects of the proposed mechanism which suggest additional experiments. Ion-pair formation, aided by a nonpolar solvent and hindered by a polar solvent, and proton exchange, influenced greatly by the presence of acids or bases in the solution, are integral parts of the mechanism. There should be some interesting solvent effects on one or more of the microscopic rate constants. The addition of an acid or a base might well have a significant effect.

BIBLIOGRAPHICAL NOTE

This reaction can be investigated by other experimental techniques, and the rate curve can be interpreted with the use of other mathematical methods. Details of laboratory procedures and of the mathematics of data treatment are given by J. F. Bunnett and H. D. Crockford [*J. Chem. Educ.*, **33**, 552 (1956)]. In particular, they discuss the reaction under conditions in which the constant-concentration approximation does not apply.

[2] J. F. Bunnett, E. W. Garbisch, Jr., and K. M. Pruitt, *J. Amer. Chem. Soc.*, **79**, 385 (1957).

8

First-Order Methods for Bimolecular Mechanisms

A great many chemical reactions are the result of the collision of two reactant molecules. The microscopic reaction in such cases involves an elementary bimolecular process. But even though that microscopic process is bimolecular, the observed macroscopic rate of reaction may be first order. This chapter deals with three experimental situations in which bimolecular reactions are found to be pseudo first order. The experimental design in each case makes possible mathematical approximations that transform unwieldy nonlinear differential equations into familiar linear differential equations of the type we have already solved. We shall call the approximations that linearize the differential equations *first-order approximations*, because the resulting approximate differential equations yield solutions in the form of first-order relaxation spectra.

This chapter deals with the constant-concentration approximation, the near-equilibrium approximation, and the equilibrium isotope tracer approximation. These approximations, when applied to rather complicated bimolecular mechanisms, allow the chemist to determine the values of many or all of the rate constants of the mechanism from experiments interpreted by integrated rate equations. The emphasis of this chapter is on ways of simplifying the mathematics by thoughtful design of experiments, thereby making analysis of data easier.

Constant-Concentration Approximation

It is often possible to plan an experiment with the concentration of one chemical species in a multireactant chemical reaction so much greater than the concentrations of the other species in solution that the relative change in this large concentration is insignificantly small during the reaction. Thus in the reaction

$$A + B = C \tag{8.1}$$

it may be possible to have (A) so large, and (B) and (C) so small, that throughout the reaction

$$(A) \simeq (A)_0 \gg (B) \tag{8.2}$$

$$(A) \simeq (A)_0 \gg (C) \tag{8.3}$$

The subscript zero denotes zero time or initial time. The quantity $(A)_0$ is, by definition, time invariant. Even complete reaction via (8.1) of all B to form C, or of all C to form B, would result in only a small percentage change in (A). Thus it is safe to assume that the instantaneous value of (A) is always essentially equal to $(A)_0$.

Assume that Reaction (8.1) takes place by means of the mechanism

$$A + B \underset{k_{-1}}{\overset{k_1}{\rightleftharpoons}} C \tag{8.4}$$

Three differential equations[1] can be written:

$$\frac{d(A)}{dt} = -k_1(A)(B) + k_{-1}(C) \tag{8.5}$$

$$\frac{d(B)}{dt} = -k_1(A)(B) + k_{-1}(C) \tag{8.6}$$

$$\frac{d(C)}{dt} = k_1(A)(B) - k_{-1}(C) \tag{8.7}$$

---

[1] The algebraic term $k_1(A)(B)$ is called a second-order term because it contains two concentration factors.

The rate of formation of C via the bimolecular elementary process

$$A + B \xrightarrow{k_1} C$$

is written as the product $k_1(A)(B)$, because it is necessary that there be an encounter between a molecule of A and a molecule of B in order that a reaction occur, and this requirement of an encounter for each molecular transformation in turn demands that

> rate of reaction $\propto$ number of encounters per second
>
> number of encounters per second $\propto$ (A)
>
> number of encounters per second $\propto$ (B)

The proportionality constant is the bimolecular rate constant, k_1, with units of seconds^{-1} [moles/liter]$^{-1}$.

Stoichiometry of the elementary reaction (8.4) requires that

$$\frac{d(A)}{dt} = \frac{d(B)}{dt} \tag{8.8}$$

and so Equations (8.5) and (8.6) are equivalent. We plan to introduce the approximation (A) $\simeq$ (A)$_0$, and so we have no interest in the time dependence of (A) during the reaction. Nothing of value to us is lost by ignoring Equation (8.5); omission of this differential equation will save time and paper in carrying through the algebra.

There is an apparent contradiction among Equations (8.2), (8.3), and (8.8). Loosely stated, Equations (8.2) and (8.3) say that (A) is constant in time, while Equation (8.8) says that (A) varies just as much as does the variable whose time dependence is being studied. More precisely, however, Equations (8.2) and (8.3) state that *relative* changes in the value of (A) are insignificantly small; *absolute* changes in (A) are equal to absolute changes in (B), according to (8.8). Even though absolute changes $\Delta(A)$ and $\Delta(B)$ are equal, the relative quantity $\Delta(A)/(A)_0$ is small compared to unity, and the quantity $\Delta(B)/(B)_0$ may be virtually equal to unity.

Equations (8.5), (8.6), and (8.7) are *nonlinear* differential equations. Terms appear with two concentration factors, and if exponential trial solutions are substituted directly into these nonlinear equations, we shall be disappointed with the results. The exponential factors will not all cancel. The task before us is to linearize the differential equations.

The constant-concentration approximation permits us to assume that the value of (A) is time independent, thus eliminating the problem of the two time-dependent concentrations in a single term. The two resulting linear differential equations[2] are

$$\frac{d(B)}{dt} = -k_1(A)_0(B) + k_{-1}(C) \qquad (8.9)$$

$$\frac{d(C)}{dt} = k_1(A)_0(B) - k_{-1}(C) \qquad (8.10)$$

Two exponential trial solutions are now assumed:

$$(B) = be^{-mt} \qquad (C) = ce^{-mt} \qquad (8.11)$$

Substitution of the trial solutions and their time derivatives into Equations (8.9) and (8.10), followed by cancellation of the identical exponential factor appearing in each term of resulting equations, leaves the two algebraic equations

$$b[k_1(A)_0 - m] - c[k_{-1}] \qquad = 0 \qquad (8.12)$$

$$-b[k_1(A)_0] \qquad + c[k_{-1} - m] = 0 \qquad (8.13)$$

Solutions of simultaneous equations (8.12) and (8.13) are the roots of the determinantal equation

$$\begin{vmatrix} [k_1(A)_0 - m] & -[k_{-1}] \\ -[k_1(A)_0] & [k_{-1} - m] \end{vmatrix} = 0 \qquad (8.14)$$

Evaluation of the determinant gives

$$[k_1(A)_0 - m][k_{-1} - m] - k_1 k_{-1}(A)_0 = 0 \qquad (8.15)$$

which simplifies to

$$m^2 - [k_{-1} + k_1(A)_0]m = 0 \qquad (8.16)$$

There are two roots to Equation (8.16):

$$m_0 = 0 \qquad (8.17)$$

$$m_1 = k_{-1} + k_1(A)_0 \qquad (8.18)$$

[2] The algebraic term $k_1(A)_0(B)$ is called a pseudo-first-order term because, although it contains two concentration factors, only one concentration is time dependent. The term behaves as if it were a first-order term.

Substitution of the roots into Equations (8.11) gives four different particular solutions, and, as is usual, none of the particular solutions is chemically meaningful by itself. The appropriate linear combinations of the particular solutions are formed, yielding two general solutions. They are

$$(B) = b_0 + b_1 e^{-m_1 t} \qquad (8.19)$$

$$(C) = c_0 + c_1 e^{-m_1 t} \qquad (8.20)$$

Experimental Determination of the Microscopic Rate Constants. Experimental determination of the macroscopic rate constant m_1 is straightforward. The chemical system approaches equilibrium via a first-order relaxation, and the numerical value of m_1 can be obtained from a rate experiment by use of the infinite-time, Guggenheim, time-lag, or half-life methods described in Chapter 3. For a single experiment, the value of m_1 yields only the sum of the microscopic rate constants. But since m_1 is concentration dependent, it is possible to evaluate both k's from data giving m_1 as a function of $(A)_0$.

If a series of experiments is performed throughout a range of initial concentrations of A, a different value for m_1 will be obtained for each value of $(A)_0$. A set of ordered pairs $\{m_1, (A)_0\}$ results from this series of experiments. A plot is then made of m_1 versus $(A)_0$. If the assumed mechanism (8.4) is consistent with these experimental data, the plot will be a straight line, the graph of Equation (8.18). The slope of this line is k_1, and the intercept at $(A)_0 = 0$ is equal to k_{-1}. This graphical method is illustrated in Figure 8.1.

The numerical value of each of the microscopic rate constants of the mechanism can thus be determined from a series of rate experiments performed at different values of $(A)_0$. More detailed information about the individual rate constants can be obtained from this bimolecular mechanism than from the simpler unimolecular mechanism which yielded only the sum of rate constants from the integrated equations.

There is an optimum range of $(A)_0$ values for experiments designed for determination of reliable values of both k_1 and k_{-1}. To get a reliable value of the intercept and thus of k_{-1}, it is necessary to have experimental data at concentrations of A such that $k_{-1} > k_1(A)_0$. To get reliable data for evaluation of the slope of the line, it is necessary to have data from experiments in which $k_{-1} < k_1(A)_0$. It is not always possible to design experiments with this ideal range of concentrations, because there are other restrictions on $(A)_0$. The upper limit on $(A)_0$ is imposed by such facts of chemical reality as (1) the concentration of a solute is limited by its solubility, and (2) no liquid can have a concentration greater than that of the pure liquid. A lower limit on $(A)_0$ is imposed by limitations of the experimental methods of detecting changes in (B) and (C), and the two inequalities (8.2) and (8.3) required for the constant-concentration approximation.

> The differential equations for a mechanism with a bimolecular elementary process contain terms with two concentration factors. These second-order terms introduce nonlinearities in the differential equations.
>
> The differential equations are linearized if the concentration of one of the reactants in the bimolecular reaction is time invariant.
>
> Time invariance of the concentration (A) is assured if experimental conditions are planned so that $(A)_0$ is so large that the chemical elementary process being studied can make no significant change in the concentration.
>
> Data from several kinetic runs performed at various values of $(A)_0$ permit straightforward numerical evaluation of both microscopic rate constants for a mechanism consisting of a single elementary reaction.

The constant-concentration approximation can be applied to the analysis of rather complicated reaction mechanisms. A reaction mechanism often is composed of elementary reactions consisting of some unimolecular processes and some bimolecular processes. Each bimolecular process introduces a second-order term into the differential equations. If an experiment can be designed so that each such second-order term contains one concentration whose value is essentially time invariant during the reaction, then the differential equations are linearized for these special experimental conditions. Example A and several of the problems at the end of this chapter are illustrations of how the constant-concentration approximation can be applied to mechanisms with several elementary reactions.

Constant-Concentration Approximation for a Buffered Species. It is often possible to arrange experimental conditions so that a chemical species, although present in very low concentration relative to concentrations of other reactants, is kept at essentially constant concentration by buffering equilibria. This use of buffering to maintain a concentration constant is commonly employed when H^+ is a reactant.

An ideal proton buffer system is composed of the species HC, C^-, and H^+, where HC and C^- do not interact chemically with any reactants except H^+. The reaction

$$HC = C^- + H^+ \tag{8.21}$$

has the equilibrium constant

$$K' = \frac{(\overline{C^-})\,(\overline{H^+})}{(\overline{HC})} \tag{8.22}$$

The acid HC should be chosen so that K' has a numerical value about equal

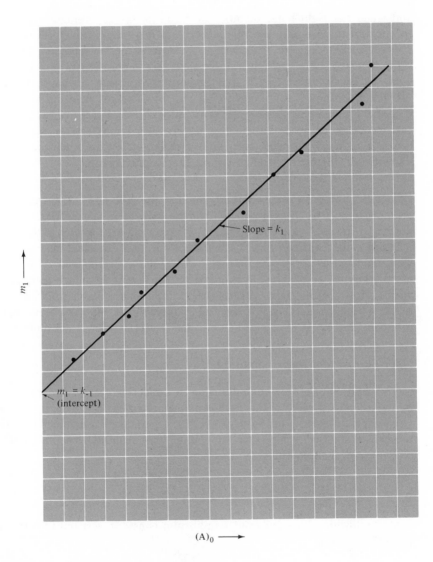

FIGURE 8.1.   Graphical evaluation of two rate constants.

to the desired value of (H^+). The concentrations (HC) and (C^-) should be approximately equal, each large enough so that the reaction under study can proceed without altering either concentration appreciably. Since

$$(H^+) = \frac{K'(\overline{HC})}{(\overline{C^-})} \tag{8.23}$$

and since the right-hand member of Equation (8.23) remains constant throughout the reaction, the value of (H^+) is necessarily time independent. It is clear that the rate of Reaction (8.21) must be fast compared to the rate processes being studied in order that (H^+) can be maintained at the value set by Equation (8.23). It is sometimes said that the concentration (H^+) is maintained at its constant value by a "rapid equilibrium."

If the numerical value of (H^+) is buffered at a time-invariant value, then each second-order term involving H^+ as a bimolecular reactant becomes a pseudo first-order term in the differential rate equations. Nonlinearities in the differential equations are thereby removed.

Constant-Concentration Approximation for the Solvent. A great many chemical reactions in solution involve the solvent as a reactant. Almost always the concentration of solvent is so large compared to any of the other reactant concentrations that it can be considered to be time independent. The situation is different, however, from that encountered with most other reactants in that the concentration of solvent ordinarily cannot be varied without drastically affecting the overall environment in which the reaction takes place. Thus it is not usually possible to evaluate a bimolecular rate constant by plotting the value of an m versus solvent concentration.

Near-Equilibrium Approximation

The position of chemical equilibrium in a reactive system is determined by the numerical value of the equilibrium constant, K, or by a set of values of equilibrium constants, $K_1, K_2, \ldots, K_n$. These K's have values that depend on temperature, pressure, and other experimental variables. Let us consider a system in which the reaction

$$A + B = C$$

is at equilibrium. The apparent equilibrium constant, written in terms of concentrations, is

$$K' = \frac{(\overline{C})}{(\overline{A})(\overline{B})}$$

where, as usual, the bars over the concentrations indicate equilibrium values.

Now suppose that the temperature of the system were changed very rapidly, rapidly compared to rates of the chemical mechanism of the reaction. The numerical value of K', being a function of temperature, will change to a new value, and there is thus established a new set of concentrations which would constitute equilibrium. The instantaneous values of the actual concentrations are not these new equilibrium values. Without themselves changing, the individual concentrations of A, B, and C have been displaced from their equilibrium values by the very fast temperature jump.

The instantaneous concentrations (A), (B), and (C) can be written as

$$(A) = \Delta(A) + (\overline{A})$$

$$(B) = \Delta(B) + (\overline{B})$$

$$(C) = \Delta(C) + (\overline{C})$$

where $\Delta(A)$, $\Delta(B)$, and $\Delta(C)$ are the displacements from equilibrium. If these displacements are small enough, the system will relax to equilibrium via first-order processes, and quantitative treatment of the rate data can proceed by means of the customary logarithmic or time-lag plots.

Experimental techniques have been developed whereby the temperature of a small sample of solution (perhaps 1 ml) can be raised 5°C in less than 10^{-7} second. This temperature jump is brought about, for an electrically conducting solution, by passing a large quantity of electricity through the sample. This large and fast input of electrical energy into the solution can be accomplished by discharging a high-voltage capacitor through the sample cell. After this initial heating, the temperature can be kept constant for a time as long as a second.

Virtually any process that perturbs the position of equilibrium very rapidly can be used to produce a displacement from equilibrium and thus permit observation of chemical relaxation back to equilibrium. Temperature-jump and pressure-jump techniques have become technically well developed. Rapid input of light or sound energy has also been widely employed. The relaxation toward chemical equilibrium can be monitored by such techniques as spectrophotometry, polarimetry, and conductrimetry, with the very fast changes in absorbance, optical rotation, or conductance displayed on the cathode-ray tube of an oscilloscope.

The approach to equilibrium of any chemical system is first order, no matter what the mechanism, if the system is sufficiently close to equilibrium. The *near-equilibrium approximation*, applied to the differential rate equations arising from mechanisms involving bimolecular elementary processes or even elementary processes of higher molecularity, linearizes those differential equations. The way this near-equilibrium approximation operates to linearize the differential equations will be illustrated with a reaction mechanism that consists of a single bimolecular forward process which is opposed by a unimolecular reverse process. The mechanistic chemical equation for the single elementary reaction is

$$A + B \underset{k_{-1}}{\overset{k_1}{\rightleftharpoons}} C \tag{8.24}$$

The associated differential rate equations are

$$\frac{d(A)}{dt} = -k_1(A)(B) + k_{-1}(C) \tag{8.25}$$

$$\frac{d(B)}{dt} = -k_1(A)(B) + k_{-1}(C) \tag{8.26}$$

$$\frac{d(C)}{dt} = k_1(A)(B) - k_{-1}(C) \tag{8.27}$$

It is convenient for purposes of the following derivation to change variables, writing Equations (8.25), (8.26), and (8.27) in terms of displacements from equilibrium. Later the assumption will be made that the reaction has approached close enough to equilibrium so that these displacements are individually quite small. These new variables are

$$(A) = (\overline{A}) + \Delta(A) \tag{8.28}$$

$$(B) = (\overline{B}) + \Delta(B) \tag{8.29}$$

$$(C) = (\overline{C}) + \Delta(C) \tag{8.30}$$

where (A), (B), and (C) are, as usual, instantaneous molar concentrations of species A, B, and C; $(\overline{A})$, $(\overline{B})$, and $(\overline{C})$ are the corresponding equilibrium concentrations; and $\Delta(A)$, $\Delta(B)$, and $\Delta(C)$ are instantaneous displacements from equilibrium. No approximation is made in writing Equations (8.28), (8.29), and (8.30). These equations serve to define the quantities $\Delta(A)$, $\Delta(B)$, and $\Delta(C)$ each of which changes with time as the chemical reaction proceeds.

Substitution of Equations (8.28)–(8.30) into Equations (8.25)–(8.27) gives a set of three equations in which the time-dependent variables are displacements from equilibrium. Certain terms will be indicated by the reference marks † and §; each of these marked terms disappears from the equations for reasons discussed below. The change of variables gives

$$\frac{d(A)}{dt} = \frac{d\Delta(A)}{dt} = -k_1[(\overline{A}) + \Delta(A)][(\overline{B}) + \Delta(B)] + k_{-1}[(\overline{C}) + \Delta(C)]$$

$$= -k_1(\overline{A})(\overline{B}) - k_1(\overline{A})\Delta(B) - k_1\Delta(A)(\overline{B}) - k_1\Delta(A)\Delta(B)$$

$$+ k_{-1}(\overline{C}) + k_{-1}\Delta(C)$$

$$= -k_1(\overline{A})(\overline{B})\dagger + k_{-1}(\overline{C})\dagger - k_1(\overline{A})\Delta(B) - k_1(\overline{B})\Delta(A)$$

$$+ k_{-1}\Delta(C) - k_1\Delta(A)\Delta(B)\S \tag{8.31}$$

$$\frac{d\Delta(B)}{dt} = -k_1(\overline{A})(\overline{B})\dagger + k_{-1}(\overline{C})\dagger - k_1(\overline{A})\Delta(B) - k_1(\overline{B})\Delta(A)$$

$$+ k_{-1}\Delta(C) - k_1\Delta(A)\Delta(B)\S \tag{8.32}$$

$$\frac{d\Delta(C)}{dt} = k_1(\overline{A})(\overline{B})\dagger - k_{-1}(\overline{C})\dagger + k_1(\overline{A})\Delta(B) + k_1(\overline{B})\Delta(A)$$

$$- k_{-1}\Delta(C) + k_1\Delta(A)\Delta(B)\S \tag{8.33}$$

In each equation the terms designated with † vanish. This fact can be seen by examining the equation at equilibrium, when the value of each concentration is equal to its equilibrium value, and when thus by definition the displacements $\Delta(A)$, $\Delta(B)$, and $\Delta(C)$ are individually equal to zero. The derivatives $d\Delta(A)/dt$, ... are each zero. Looking then at Equation (8.31), we note that most of the terms become zero at equilibrium, and there is left only

$$0 = -k_1(\overline{A})(\overline{B}) + k_{-1}(\overline{C}) \tag{8.34}$$

Since all quantities in Equation (8.34) are time independent, Equation (8.34) is valid not only at equilibrium but also throughout the entire reaction. This pair of terms can be eliminated *without approximation* from Equation (8.31). In the same manner, the terms marked with † vanish without approximation from Equations (8.32) and (8.33).

The near-equilibrium chemical approximation enters the mathematics at this point, for here the terms marked with § are to be discarded. This is clearly an approximation, and the approximation would not be expected to be generally valid. However, if the reaction system is in the neighborhood of equilibrium so that the concentrations are close enough to their respective equilibrium values, then $\Delta(A)$ and $\Delta(B)$ will individually be small enough so that their product $\Delta(A)\Delta(B)$ is insignificantly small with respect to either $(\overline{A})\Delta(B)$ or $(\overline{B})\Delta(A)$. The product $\Delta(A)\Delta(B)$ becomes smaller and smaller, and therefore the approximation becomes better and better, as the reacting system evolves toward equilibrium. Written in a more explicit fashion, the approximation is either

$$\Delta(A)\Delta(B) \ll (\overline{A})\Delta(B)$$

or

$$\Delta(A)\Delta(B) \ll (\overline{B})\Delta(A)$$

These inequalities are equivalent to either

$$\Delta(A) \ll (\overline{A})$$

or

$$\Delta(B) \ll (\overline{B})$$

    With all terms marked with † or § eliminated, Equations (8.31)–(8.33) become

$$\frac{d\Delta(A)}{dt} = -[k_1(\overline{B})]\Delta(A) - [k_1(\overline{A})]\Delta(B) + k_{-1}\Delta(C) \qquad (8.35)$$

$$\frac{d\Delta(B)}{dt} = -[k_1(\overline{B})]\Delta(A) - [k_1(\overline{A})]\Delta(B) + k_{-1}\Delta(C) \qquad (8.36)$$

$$\frac{d\Delta(C)}{dt} = [k_1(\overline{B})]\Delta(A) + [k_1(\overline{A})]\Delta(B) - k_{-1}\Delta(C) \qquad (8.37)$$

The exact differential equations for the mechanism are nonlinear. Introduction of the near-equilibrium approximation, together with the change of variables from concentrations to concentration differences, linearizes these differential equations. The same mathematical methods that we have used several times previously can now be employed again to obtain solutions to this set of simultaneous, first-order, linear differential equations.

    Solutions are now assumed of the form

$$\Delta(A) = ae^{-mt} \qquad (8.38)$$

$$\Delta(B) = be^{-mt} \qquad (8.39)$$

$$\Delta(C) = ce^{-mt} \qquad (8.40)$$

Substitution of assumed solutions (8.38), (8.39), and (8.40) into differential equations (8.35), (8.36), and (8.37) yields three algebraic equations.

Each term in each equation contains the common factor e^{-mt}. Division by this exponential leaves

$$a[k_1(\overline{B}) - m] + b[k_1(\overline{A})] \qquad - c[k_{-1}] \qquad = 0 \qquad (8.41)$$

$$a[k_1(\overline{B})] \qquad + b[k_1(\overline{A}) - m] - c[k_{-1}] \qquad = 0 \qquad (8.42)$$

$$-a[k_1(\overline{B})] \qquad - b[k_1(\overline{A})] \qquad + c[k_{-1} - m] = 0 \qquad (8.43)$$

This set of three simultaneous linear equations can be solved by finding the three roots of the determinantal equation

$$\begin{vmatrix} [k_1(\overline{B}) - m] & [k_1(\overline{A})] & -[k_{-1}] \\ [k_1(\overline{B})] & [k_1(\overline{A}) - m] & -[k_{-1}] \\ -[k_1(\overline{B})] & -[k_1(\overline{A})] & [k_{-1} - m] \end{vmatrix} = 0 \qquad (8.44)$$

Evaluation of the determinant gives

$$m^3 - m^2\{k_1[(\overline{A}) + (\overline{B})] + k_{-1}\} = 0 \qquad (8.45)$$

There are three roots to this cubic equation, but two are identical and equal to zero. Nothing is to be gained by distinguishing between the two identical roots, and so the roots will be labeled as

$$m_0 = 0 \qquad (8.46)$$

$$m_1 = k_1[(\overline{A}) + (\overline{B})] + k_{-1} \qquad (8.47)$$

Substitution of Equations (8.46) and (8.47) into the assumed solution (8.38) yields two particular solutions, and then formation of the linear combination of the particular solutions yields the general solution

$$\Delta(A) = a_0 + a_1 e^{-m_1 t} \qquad (8.48)$$

Variables can be changed back again with the aid of Equation (8.28), giving

$$\Delta(A) = (A) - (\overline{A}) = a_0 + a_1 e^{-m_1 t}$$

$$(A) = [a_0 + (\overline{A})] + a_1 e^{-m_1 t} \qquad (8.49)$$

Each of the three concentrations is also a similar function of time. The value of m_1 can be determined from experimental data by use of the various first-order methods which have been discussed in Chapter 3.

Because the macroscopic rate constant m_1 is a linear function of the equilibrium concentrations, both microscopic rate constants can be evaluated from the results of a series of rate experiments conducted over a range of concentrations of reactants. Then a plot of m_1 versus the sum $[(\overline{A}) + (\overline{B})]$ should be a straight line with slope of k_1 and intercept of k_{-1}.

Linearization of the differential equations is achieved by assuming that displacements of the concentrations from their equilibrium values are individually very small. The derivation proceeds as follows:

The time-dependent variables are changed from concentrations to corresponding displacements of the concentrations from their equilibrium values.

In the differential equations resulting from the change of variables, certain terms involving equilibrium concentrations are found to sum to zero. These terms are eliminated as a group from each equation. But the differential equations remain nonlinear.

Linearization is achieved by eliminating each term containing more than one displacement factor. This mathematical operation is justified if the experimental condition is met that the reaction is very close to equilibrium.

There is one elementary reaction, to which there correspond two microscopic rate constants and one macroscopic rate constant. The numerical value of the first-order macroscopic rate constant depends on the equilibrium concentrations, and thus it is possible to evaluate both the bimolecular rate constant k_1 and the unimolecular rate constant k_{-1} from experimental data obtained from an appropriate series of experiments run at different reactant concentrations.

This method is completely general. For any mechanism, no matter how complicated, similar mathematics yields linear differential equations which in turn predict first-order relaxations, provided only that the displacements from equilibrium are small enough.

Equilibrium Isotope Tracer Approximation

Another powerful method of investigating chemical reaction mechanisms takes advantage of isotopically labeled reactants, reactants in which one or more atoms have been replaced by a different isotope of the same element. Especially if the isotope used is radioactive, it is possible to detect the labeled molecules in very low concentrations, and to make quantitative measurements of those concentrations. The labeled molecules act as tracers moving through the reaction sequence. Experimental conditions can be planned so that the observed rates of isotope incorporation in any of the intermediates or products will be first-order relaxations for a great many mechanisms. If the initial concentration of the isotopically labeled reactant is very small compared to the corresponding unlabeled compound, and if chemical equilibrium has been established in the system prior to introduction of the tracer, then for even quite complicated reaction mechanisms it will be found that the system will approach isotopic-distribution equilibrium via first-order relaxations.

Any mechanism in which the molecularity with respect to labeled species is unity in each elementary reaction will yield first-order rates provided only that the concentrations of unlabeled species do not change appreciably during the course of the isotopic-exchange reactions. Two conditions under the control of the experimental chemist will suffice to keep these concentrations constant: (1) the initial concentration of the tracer compound must be much smaller than the equilibrium concentration of its unlabeled counterpart, and (2) chemical equilibrium must be established in the system before the introduction of the tracer.

As an example, consider the reaction

$$AX + B = A + BX \tag{8.50}$$

studied by introducing AX* (compound AX* is AX with isotopic substitution of an atom in the X group) into an equilibrium mixture of AX, A, BX, and B. Suppose that the reaction mechanism followed is

$$AX + B \underset{k_{-1}}{\overset{k_1}{\rightleftharpoons}} A + BX \tag{8.51}$$

Then the mechanism for distribution of the isotopic label among the reactants and products is

$$AX^* + B \underset{k_{-1}}{\overset{k_1}{\rightleftharpoons}} A + BX^* \tag{8.52}$$

The two differential rate equations associated with Reaction (8.52) are

$$\frac{d(AX^*)}{dt} = -k_1(AX^*)(\overline{B}) + k_{-1}(\overline{A})(BX^*) \tag{8.53}$$

$$\frac{d(BX^*)}{dt} = k_1(AX^*)(\overline{B}) - k_{-1}(\overline{A})(BX^*) \tag{8.54}$$

Experimental conditions have been so arranged that throughout the reaction the concentrations $(\overline{A})$ and $(\overline{B})$ are time invariant. Thus in spite of the fact that the elementary reaction consists of two bimolecular processes, the differential equations are linear. We may confidently predict first-order behavior.

Solutions of the form

$$(AX^*) = ae^{-mt} \tag{8.55}$$

$$(BX^*) = be^{-mt} \tag{8.56}$$

are introduced. The exponentials that appear in each term are canceled, with the result

$$a[k_1(\overline{B}) - m] - b[k_{-1}(\overline{A})] = 0 \tag{8.57}$$

$$-a[k_1(\overline{B})] + b[k_{-1}(\overline{A}) - m] = 0 \tag{8.58}$$

For arbitrary values of the parameters a and b, Equations (8.57) and (8.58) are simultaneously satisfied by either of two values of m:

$$m_0 = 0 \tag{8.59}$$

$$m_1 = k_1(\overline{B}) + k_{-1}(\overline{A}) \tag{8.60}$$

The chemically meaningful, general solutions of the differential equations are

$$(AX^*) = a_0 + a_1 e^{-m_1 t} \tag{8.61}$$

$$(BX^*) = b_0 + b_1 e^{-m_1 t} \tag{8.62}$$

The concentration of each of the isotopically labeled chemical species changes with time via a first-order relaxation process characterized in each case by the same macroscopic first-order rate constant. Both bimolecular microscopic rate constants of the mechanism can be evaluated numerically from data obtained in a series of experiments in which the values of the time-independent concentrations $(\overline{A})$ and $(\overline{B})$ are varied.

In general, no matter what the molecularity of the individual elementary processes with respect to unlabeled species, the presence of a single labeled species on each side of each chemical equation in the mechanism results in a set of linear differential equations. This useful first-order simplification occurs because, in each elementary process involving a labeled molecule, all except one of the concentrations have time-invariant values. Establishment of chemical equilibrium in all the unlabeled species, and the presence of only very small concentrations of the tracer compounds, are the experimental conditions that assure this time independence of un-labeled-species concentrations.

WORKED EXAMPLES

Example A. Show how all the microscopic rate constants can be evaluated from experimental data for the chemical reaction

$$A + B = C$$

which takes place via the mechanism

$$A + B \underset{k_{-1}}{\overset{k_1}{\rightleftharpoons}} X$$

$$X \underset{k_{-2}}{\overset{k_2}{\rightleftharpoons}} C$$

Suppose that the data available were collected under conditions such that $(A)_0 \gg (B)_0$, and $(A)_0 \gg (C)_0$.

Answer. The differential rate equations are

$$\frac{d(B)}{dt} = -k_1(A)(B) + k_{-1}(X)$$

$$\frac{d(X)}{dt} = k_1(A)(B) - k_{-1}(X) - k_2(X) + k_{-2}(C)$$

$$\frac{d(C)}{dt} = k_2(X) - k_{-2}(C)$$

Introduce the constant-concentration approximation in the form $(A) = (A)_0$. Assume trial solutions of the form

$$(B) = be^{-mt} \qquad (X) = xe^{-mt} \qquad (C) = ce^{-mt}$$

Introduce these solutions into the differential equations, cancel the common exponential factor from each term in each equation, and get

$$b[k_1(A)_0 - m] - x[k_{-1}] \qquad\qquad\qquad\qquad = 0$$
$$-b[k_1(A)_0] \qquad + x[k_{-1} + k_2 - m] - c[k_{-2}] \quad = 0$$
$$\qquad\qquad\qquad - x[k_2] \qquad\qquad\qquad + c[k_{-2} - m] = 0$$

An explicit relationship between the k's and the three m's is found by solving the determinantal equation

$$\begin{vmatrix} [k_1(A)_0 - m] & -[k_{-1}] & 0 \\ -[k_1(A)_0] & [k_{-1} + k_2 - m] & -[k_{-2}] \\ 0 & -[k_2] & [k_{-2} - m] \end{vmatrix} = 0$$

Expansion of the determinant yields the equation

$$m^3 - m^2[k_1(A)_0 + k_{-1} + k_2 + k_{-2}] + m[k_1 k_{-2}(A)_0 \\ + k_{-1}k_{-2} + k_1 k_2(A)_0] = 0$$

This result is identical with Equation (5.18) except that each k_1 in Equation (5.18) has been replaced by $k_1(A)_0$. This difference, however, is critical, for it provides the kineticist with a means for evaluating each of the microscopic rate constants. The nonzero roots of the cubic equation are

$$m_1 + m_2 = k_1(A)_0 + [k_{-1} + k_2 + k_{-2}]$$
$$m_1 m_2 = (A)_0 k_1 [k_2 + k_{-2}] + k_{-1}k_{-2}$$

Thus a plot of $[m_1 + m_2]$ versus $(A)_0$ yields k_1 directly as the slope and the sum $[k_{-1} + k_2 + k_{-2}]$ as the intercept. A plot of $m_1 m_2$ gives $k_1[k_2 + k_{-2}]$ and, since k_1 is now known, the sum $[k_2 + k_{-2}]$ from the slope. The intercept of the second plot is $k_{-1}k_{-2}$. Subtraction of $[k_2 + k_{-2}]$ from $[k_{-1} + k_2 + k_{-2}]$ gives k_{-1}. Division of $k_{-1}k_{-2}$ by k_{-1} gives k_{-2}. Finally, subtraction of k_{-2} from $[k_2 + k_{-2}]$ gives k_2. This method is sometimes limited in application where the numerical values of the various microscopic rate constants differ widely in value.

Example B.   Consider a kinetics experiment performed by displacing a reactive system slightly from equilibrium, perhaps by means of a very rapid change in temperature or pressure. The reaction mechanism is

$$A + B \underset{k_{-1}}{\overset{k_1}{\rightleftharpoons}} X$$

$$X \underset{k_{-2}}{\overset{k_2}{\rightleftharpoons}} C + D$$

Find relationships between the k's of the mechanism and the first-order macroscopic rate constants obtained from the experiment.

Answer.   The differential rate equations are

$$\frac{d(\mathrm{A})}{dt} = -k_1(\mathrm{A})(\mathrm{B}) + k_{-1}(\mathrm{X})$$

$$\frac{d(\mathrm{B})}{dt} = -k_1(\mathrm{A})(\mathrm{B}) + k_{-1}(\mathrm{X})$$

$$\frac{d(\mathrm{X})}{dt} = k_1(\mathrm{A})(\mathrm{B}) - k_{-1}(\mathrm{X}) - k_2(\mathrm{X}) + k_{-2}(\mathrm{C})(\mathrm{D})$$

$$\frac{d(\mathrm{C})}{dt} = k_2(\mathrm{X}) - k_{-2}(\mathrm{C})(\mathrm{D})$$

$$\frac{d(\mathrm{D})}{dt} = k_2(\mathrm{X}) - k_{-2}(\mathrm{C})(\mathrm{D})$$

Change variables according to

$$(\mathrm{A}) = (\overline{\mathrm{A}}) + \Delta(\mathrm{A})$$
$$(\mathrm{B}) = (\overline{\mathrm{B}}) + \Delta(\mathrm{B})$$
$$(\mathrm{X}) = (\overline{\mathrm{X}}) + \Delta(\mathrm{X})$$
$$(\mathrm{C}) = (\overline{\mathrm{C}}) + \Delta(\mathrm{C})$$
$$(\mathrm{D}) = (\overline{\mathrm{D}}) + \Delta(\mathrm{D})$$

In terms of the displacement variables, the differential equations are

$$\frac{d\Delta(\mathrm{A})}{dt} = -k_1[(\overline{\mathrm{A}}) + \Delta(\mathrm{A})][(\overline{\mathrm{B}}) + \Delta(\mathrm{B})] + k_{-1}[(\overline{\mathrm{X}}) + \Delta(\mathrm{X})]$$

$$\frac{d\Delta(\mathrm{B})}{dt} = -k_1[(\overline{\mathrm{A}}) + \Delta(\mathrm{A})][(\overline{\mathrm{B}}) + \Delta(\mathrm{B})] + k_{-1}[(\overline{\mathrm{X}}) + \Delta(\mathrm{X})]$$

$$\frac{d\Delta(\mathrm{X})}{dt} = k_1[(\overline{\mathrm{A}}) + \Delta(\mathrm{A})][(\overline{\mathrm{B}}) + \Delta(\mathrm{B})] - [k_{-1} + k_2][(\overline{\mathrm{X}}) + \Delta(\mathrm{X})]$$
$$+ k_{-2}[(\overline{\mathrm{C}}) + \Delta(\mathrm{C})][(\overline{\mathrm{D}}) + \Delta(\mathrm{D})]$$

$$\frac{d\Delta(\mathrm{C})}{dt} = k_2[(\overline{\mathrm{X}}) + \Delta(\mathrm{X})] - k_{-2}[(\overline{\mathrm{C}}) + \Delta(\mathrm{C})][(\overline{\mathrm{D}}) + \Delta(\mathrm{D})]$$

$$\frac{d\Delta(\mathrm{D})}{dt} = k_2[(\overline{\mathrm{X}}) + \Delta(\mathrm{X})] - k_{-2}[(\overline{\mathrm{C}}) + \Delta(\mathrm{C})][(\overline{\mathrm{D}}) + \Delta(\mathrm{D})]$$

The indicated multiplications are performed. Use is made of the relationships

$$k_1(\overline{A})(\overline{B}) = k_{-1}(\overline{X})$$

$$k_{-2}(\overline{C})(\overline{D}) = k_2(\overline{X})$$

to eliminate certain terms from the differential equations. The near-equilibrium approximation is introduced to eliminate all terms containing the product of two displacements. There is left

$$\frac{d\Delta(A)}{dt} = -k_1(\overline{B})\Delta(A) - k_1(\overline{A})\Delta(B) + k_{-1}\Delta(X)$$

$$\frac{d\Delta(B)}{dt} = -k_1(\overline{B})\Delta(A) - k_1(\overline{A})\Delta(B) + k_{-1}\Delta(X)$$

$$\frac{d\Delta(X)}{dt} = k_1(\overline{B})\Delta(A) + k_1(\overline{A})\Delta(B) - k_{-1}\Delta(X) - k_2\Delta(X) + k_{-2}(\overline{C})\Delta(D)$$

$$+ k_{-2}(\overline{D})\Delta(C)$$

$$\frac{d\Delta(C)}{dt} = k_2\Delta(X) - k_{-2}(\overline{C})\Delta(D) - k_{-2}(\overline{D})\Delta(C)$$

$$\frac{d\Delta(D)}{dt} = k_2\Delta(X) - k_{-2}(\overline{C})\Delta(D) - k_{-2}(\overline{D})\Delta(C)$$

Try solutions of the form

$$\Delta(A) = ae^{-mt} \qquad \Delta(B) = be^{-mt} \qquad \Delta(X) = xe^{-mt}$$

$$\Delta(C) = ce^{-mt} \qquad \Delta(D) = de^{-mt}$$

Substitution of these trial solutions into the differential equations yields, after cancellation of the common exponential factor,

$$a[k_1(\overline{B}) - m] + b[k_1(\overline{A})] - x[k_{-1}] = 0$$

$$a[k_1(\overline{B})] + b[k_1(\overline{A}) - m] - x[k_{-1}] = 0$$

$$-a[k_1(\overline{B})] - b[k_1(\overline{A})] + x[k_{-1} + k_2 - m]$$

$$- c[k_{-2}(\overline{D})] - d[k_{-2}(\overline{C})] = 0$$

$$-x[k_2] + c[k_{-2}(\overline{D}) - m] + d[k_{-2}(\overline{C})] = 0$$

$$-x[k_2] + c[k_{-2}(\overline{D})] + d[k_{-2}(\overline{C}) - m] = 0$$

The required relationship between the k's and the experimentally observable m's can be found by solving the associated determinantal equation. The five-by-five determinant takes quite a long while to expand, but the resulting quintic equation is

$$m^5 - m^4\{k_1[(\overline{A}) + (\overline{B})] + k_{-1} + k_2 + k_{-2}[(\overline{C}) + (\overline{D})]\}$$

$$+ m^3\{k_1k_2[(\overline{A}) + (\overline{B})] + k_1k_{-2}[(\overline{A}) + (\overline{B})][(\overline{C}) + (\overline{D})]$$

$$+ k_{-1}k_{-2}[(\overline{C}) + (\overline{D})]\} = 0$$

Three roots of this equation are identical, each equal to zero. The nonzero roots are

$$m_1 + m_2 = k_1[(\overline{A}) + (\overline{B})] + k_{-1} + k_2 + k_{-2}[(\overline{C}) + (\overline{D})]$$

$$m_1 m_2 = k_1 k_2[(\overline{A}) + (\overline{B})] + k_1 k_{-2}[(\overline{A}) + (\overline{B})][(\overline{C}) + (\overline{D})]$$

$$+ k_{-1}k_{-2}[(\overline{C}) + (\overline{D})]$$

Alternative, Shortcut Answer. Much of the labor of the preceding lengthy answer is unnecessary. Because of the stoichiometry of the elementary reaction steps, it is necessary that throughout the entire approach to equilibrium

$$\Delta(B) = \Delta(A) \qquad \text{and} \qquad \Delta(C) = \Delta(D)$$

Substitution of these equations into the original differential equations reveals that there are two pairs of identical equations. All the information in the problem is contained in the three differential equations

$$\frac{d\Delta(B)}{dt} = -k_1[(\overline{B}) + (\overline{A})]\Delta(B) + k_{-1}\Delta(X)$$

$$\frac{d\Delta(X)}{dt} = k_1[(\overline{B}) + (\overline{A})]\Delta(B) - [k_{-1} + k_2]\Delta(X) + k_{-2}[(\overline{C}) + (\overline{D})]\Delta(C)$$

$$\frac{d\Delta(C)}{dt} = k_2\Delta(X) - k_{-2}[(\overline{C}) + (\overline{D})]\Delta(C)$$

Solutions are assumed of the form

$$\Delta(B) = be^{-mt} \qquad \Delta(X) = xe^{-mt} \qquad \Delta(C) = ce^{-mt}$$

Introduction of the trial solutions into the differential equations gives

$$b\{k_1[(\overline{B}) + (\overline{A})] - m\} - x[k_{-1}] = 0$$
$$-bk_1[(\overline{B}) + (\overline{A})] \quad + x[k_{-1} + k_2 - m] - ck_{-2}[(\overline{C}) + (\overline{D})] = 0$$
$$- x[k_2] \quad + c\{k_{-2}[(\overline{C}) + (\overline{D})] - m\} = 0$$

The associated determinantal equation is

$$\begin{vmatrix} \{k_1[(\overline{B}) + (\overline{A})] - m\} & -[k_{-1}] & 0 \\ -\{k_1[(\overline{B}) + (\overline{A})]\} & [k_{-1} + k_2 - m] & -\{k_{-2}[(\overline{C}) + (\overline{D})]\} \\ 0 & -[k_2] & \{k_{-2}[(\overline{C}) + (\overline{D})] - m\} \end{vmatrix} = 0$$

Expansion of the determinant yields

$$m^3 - m^2\{k_1[(\overline{A}) + (\overline{B})] + k_{-1} + k_2 + k_{-2}[(\overline{C}) + (\overline{D})]\}$$
$$+ m\{k_1 k_2[(\overline{A}) + (\overline{B})] + k_1 k_{-2}[(\overline{A}) + (\overline{B})][(\overline{C}) + (\overline{D})]$$
$$+ k_{-1} k_{-2}[(\overline{C}) + (\overline{D})]\} = 0$$

One root of this cubic equation is zero, and the two nonzero roots are identical to the roots obtained by the more lengthy procedure. Since the two sums of equilibrium concentrations are independently variable and under the control of the chemist who mixes the solutions, the concentration dependence of the m's gives enough information for the numerical evaluation of each of the microscopic rate constants of the mechanism.

Example C. The reaction

$$AX + B = A + BX$$

is allowed to come to equilibrium. Then a very small amount of AX* (AX* is the compound AX with an isotopic atom substituted in the X group) is introduced, and the rate of incorporation of label in BX is observed. Two different mechanisms might be envisioned:

Dissociative Mechanism:

$$AX \underset{k_{-1}}{\overset{k_1}{\rightleftharpoons}} A + X$$

$$B + X \underset{k_{-2}}{\overset{k_2}{\rightleftharpoons}} BX$$

Associative Mechanism:

$$AX + B \underset{k_{-1}}{\overset{k_1}{\rightleftharpoons}} AXB$$

$$AXB \underset{k_{-2}}{\overset{k_2}{\rightleftharpoons}} A + BX$$

Show the difference between these two mechanisms in terms of the dependence of the first-order macroscopic rate constants on the concentrations of the unlabeled reactants, and indicate how experimental rate data can be used to distinguish between the mechanisms.

Answer.   Since the system of unlabeled reactants and products is at equilibrium, and since the tracer compound is added in only very low concentration, we shall assume constancy of (AX), (A), (B), and (BX). The two mechanisms will be considered separately, and then the results will be compared.

Dissociative Mechanism.   The relevant differential equations are

$$\frac{d(AX^*)}{dt} = -k_1(AX^*) + k_{-1}(\overline{A})(X^*)$$

$$\frac{d(X^*)}{dt} = k_1(AX^*) - k_{-1}(\overline{A})(X^*) - k_2(\overline{B})(X^*) + k_{-2}(BX^*)$$

$$\frac{d(BX^*)}{dt} = k_2(\overline{B})(X^*) - k_{-2}(BX^*)$$

There are only three time-dependent concentrations, and for each we try an exponential solution:

$$(AX^*) = ae^{-mt} \qquad (X^*) = xe^{-mt} \qquad (BX^*) = be^{-mt}$$

Substitution of the trial solutions into the differential equations yields

$$a[k_1 - m] - x[k_{-1}(\overline{A})] \qquad\qquad\qquad = 0$$

$$-a[k_1] \quad + x[k_{-1}(\overline{A}) + k_2(\overline{B}) - m] - b[k_{-2}] \quad = 0$$

$$- x[k_2(\overline{B})] \qquad\qquad + b[k_{-2} - m] = 0$$

The relationship between the k's and the macroscopic relaxation constants

m_1 and m_2 can be found by solving the determinantal equation

$$
\begin{vmatrix}
[k_1 - m] & -[k_{-1}(\overline{A})] & 0 \\
-[k_1] & [k_{-1}(\overline{A}) + k_2(\overline{B}) - m] & -[k_{-2}] \\
0 & -[k_2(\overline{B})] & [k_{-2} - m]
\end{vmatrix} = 0
$$

Expansion of the determinant results in

$$m^3 - m^2[k_1 + k_{-1}(\overline{A}) + k_2(\overline{B}) + k_{-2}]$$
$$+ m[k_1 k_{-2} + k_{-1} k_{-2}(\overline{A}) + k_1 k_2(\overline{B})] = 0$$

The nonzero roots are

$$m_1 + m_2 = k_1 + k_{-1}(\overline{A}) + k_2(\overline{B}) + k_{-2}$$
$$m_1 m_2 = k_1 k_{-2} + k_{-1} k_{-2}(\overline{A}) + k_1 k_2(\overline{B})$$

Associative Mechanism.   The differential rate equations for this mechanism are

$$\frac{d(\text{AX*})}{dt} = -k_1(\overline{B})(\text{AX*}) + k_{-1}(\text{AX*B})$$

$$\frac{d(\text{AX*B})}{dt} = k_1(\overline{B})(\text{AX*}) - k_{-1}(\text{AX*B}) - k_2(\text{AX*B}) + k_{-2}(\overline{A})(\text{BX*})$$

$$\frac{d(\text{BX*})}{dt} = k_2(\text{AX*B}) - k_{-2}(\overline{A})(\text{BX*})$$

Substitution of trial solutions

$$(\text{AX*}) = ae^{-mt} \qquad (\text{AX*B}) = xe^{-mt} \qquad (\text{BX*}) = be^{-mt}$$

results in

$$a[k_1(\overline{B}) - m] - x[k_{-1}] = 0$$
$$-a[k_1(\overline{B})] + x[k_{-1} + k_2 - m] - b[k_{-2}(\overline{A})] = 0$$
$$- x[k_2] + b[k_{-2}(\overline{A}) - m] = 0$$

The determinantal equation is

$$
\begin{vmatrix}
[k_1(\overline{B}) - m] & -[k_{-1}] & 0 \\
-[k_1(\overline{B})] & [k_{-1} + k_2 - m] & -[k_{-2}(\overline{A})] \\
0 & -[k_2] & [k_{-2}(\overline{A}) - m]
\end{vmatrix} = 0
$$

which is equivalent to

$$
m^3 - m^2[k_1(\overline{B}) + k_{-1} + k_2 + k_{-2}(\overline{A})]
$$
$$
+ m[k_1 k_{-2}(\overline{A})(\overline{B}) + k_{-1}k_{-2}(\overline{A}) + k_1 k_2(\overline{B})] = 0
$$

The nonzero roots are

$$
m_1 + m_2 = k_1(\overline{B}) + k_{-1} + k_2 + k_{-2}(\overline{A})
$$
$$
m_1 m_2 = k_1 k_{-2}(\overline{A})(\overline{B}) + k_{-1}k_{-2}(\overline{A}) + k_1 k_2(\overline{B})
$$

Comparison. A plot of $m_1 m_2$ versus $(\overline{A})$ at constant $(\overline{B})$ will give a straight line for each mechanism, but the slopes and intercepts of the plots are different in a significant way:

Mechanism	Slope	Intercept
Dissociative	$k_{-1}k_{-2}$	$k_1 k_{-2} + k_1 k_2(\overline{B})$
Associative	$k_1 k_{-2}(\overline{B}) + k_{-1}k_{-2}$	$k_1 k_2(\overline{B})$

If experimental data are sufficiently accurate, then there is a straight-forward test to distinguish between the two mechanisms. A plot of slope versus $(\overline{B})$ has a slope of zero for the dissociative mechanism, but a slope of $k_1 k_{-2}$ for the associative mechanism. A plot of intercept versus $(\overline{B})$ has an intercept of zero for the associative mechanism, but an intercept of $k_1 k_{-2}$ for the dissociative mechanism. The chemist must be wary of the effect of cumulative errors in this procedure, for the critical test between mechanisms lies three graphing steps away from the experimental data. Plots have first to be made from experimental data for evaluation of m_1 and m_2 at particular values of $(\overline{B})$ and $(\overline{A})$. The product of the two m's is then plotted versus $(\overline{A})$ for each of several values of $(\overline{B})$. Finally, the slope and intercept of each of these $m_1 m_2$ versus $(\overline{A})$ plots is plotted versus $(\overline{B})$. The slope and intercept of this final plot are subject to very large uncertainties.

SUGGESTIONS FOR EXPERIMENTAL INVESTIGATION

Continuous-Flow Method: Reaction of Ferric and Thiocyanate Ions.   The reaction of Fe^{3+} and SCN^- to yield the complex $FeSCN^{2+}$ has a half-life of less than 1 second when $(Fe^{3+})_0$ is 10^{-2} molar and $(SCN^-)_0$ is 10^{-3} molar. The reaction is accompanied by change in the color of the solution. The Fe^{3+} is present in sufficient excess so that a first-order relaxation occurs.

This reaction is too fast for rate studies to be made by mixing the reactant solutions in a beaker and then measuring optical absorbance in a spectrophotometer. A continuous-flow apparatus, described in detail by D. P. Shoemaker and C. W. Garland [*Experiments in Physical Chemistry*, New York: McGraw-Hill, Inc., 2nd ed., 1967, experiment 30, "Kinetics of a Fast Reaction," p. 241], is well suited for measuring the rate of this reaction. Mixing can be accomplished in about 10^{-3} second, and then the reacting mixture flows down a long capillary tube. Measurements of optical absorbance can be taken at various distances along the capillary tube with a modified Beckman DU spectrophotometer. The apparatus is designed so that distance along the tube is proportional to the elapsed time after mixing. Several liters of reactant solution are required for these measurements.

Concentration Jump with Rapid Mixing: Chromate–Dichromate Interconversion.   The position of equilibrium with respect to the reaction

$$2HCrO_4^- = Cr_2O_7^{2-} + H_2O$$

is shifted when the solution is diluted with water, water being in this case both solvent and a reactant. Experimental conditions can be arranged so that relaxation toward the new equilibrium point occurs with a half-life in the range 10 to 100 seconds. J. H. Swinehart [*J. Chem. Educ.*, **44**, 524 (1967)] describes a student experiment for measurement of the first-order macroscopic rate constant for this relaxation and shows how two microscopic rate constants for a plausible mechanism can be evaluated from the data obtained. He describes a simple mixing device which permits mixing of solutions in a spectrophotometer cell in about 2 seconds. The relaxation is followed with a spectrophotometer, using an acid–base indicator to monitor the reaction of interest.

Radioactive Tracer Method: Carbonate Exchange with [Co en$_2$CO$_3$]$^+$.   The rate of the reaction

$$[Co\ en_2CO_3]^+ + C^*O_3^{2-} = [Co\ en_2C^*O_3]^+ + CO_3^{2-}$$

can be studied conveniently [D. Barton and K. Winter, *J. Chem. Educ.*, **43**, 93 (1966)] in the undergraduate laboratory. A very small quantity of carbon-14-labeled Na_2CO_3 is introduced into a solution containing the complex ion and unlabeled Na_2CO_3. At intervals during a 2-hour period, samples of the reacting solution are withdrawn and the uncomplexed carbonate precipitated with $BaCl_2$. The radioactivity of the precipitated $BaCO_3$ is measured for each sample with a Geiger–Müller counter. The infinite-time value of carbonate radioactivity is not reliable, according to Barton and Winter, and they suggest calculating it. An alternative is to take enough data so that a Guggenheim plot can be used. This requires withdrawing samples at more frequent intervals.

Propose a mechanism for this reaction (see, for instance, C. H. Langford and H. B. Gray, *Ligand Substitution Processes*, New York: W. A. Benjamin, Inc., 1965, chaps. 1 and 3), and attempt to make a quantitative statement about values of the rate constants for your mechanism in terms of the data from this ligand exchange experiment. What additional experimental data would be useful?

Constant-Concentration Approximation: Acid Catalysis of the Hydrolysis of a Lactone.   The compound γ-butyrolactone,

$$
\begin{array}{c}
H_2 \\
C \\
\diagup \quad \diagdown \\
H_2{-}C \qquad C{=}O \\
\diagdown \quad \diagup \\
C{-}O \\
H_2
\end{array}
$$

L

reacts in water to produce γ-hydroxybutyric acid,

$$
HO{-}CH_2{-}CH_2{-}CH_2{-}C \begin{array}{c} O \\ \diagup\diagup \\ \diagdown \\ OH \end{array}
$$

B

The five-membered ring is opened in the reaction. The reaction proceeds at measurable rates in the presence of acid or base, but an attempt [F. D. Coffin and F. A. Long, *J. Amer. Chem. Soc.*, **74**, 5767 (1952)] to demonstrate a reaction pathway involving neither an acid nor a base proved unsuccessful. Probably different mechanisms are operative in acidic and in basic media. We shall consider the reaction in acid solution.

A. R. Osborn and E. Whalley [*Trans. Faraday Soc.*, **58**, 2144 (1962)] followed the hydrolysis of the lactone, initially about 0.3 molar in a 0.05 molar hydrochloric acid solution, by titration with standardized base. The reaction rate curve was found to be a first-order relaxation curve. If the reaction mechanism is simply

$$L + H^+ + H_2O \underset{k_{-1}}{\overset{k_1}{\rightleftharpoons}} B + H^+$$

then the single observed macroscopic rate constant m_1 is given by

$$m_1 = k_1(H_2O)(H^+) + k_{-1}(H^+)$$

H^+ is neither consumed nor created within this mechanism, and its concentration remains constant throughout the reaction. There is an overwhelming excess of solvent compared to lactone, and so it is justifiable to assume (H_2O) constant. This mechanism predicts that a plot of m_1 versus (H^+) is linear with a zero intercept. With this mechanism, there is the particularly simple relationship between the rate constants and the equilibrium constant,

$$K' = \frac{k_1}{k_{-1}}$$

The equilibrium constant was determined experimentally by Osborn and Whalley, and they used its value, the value of m_1, and the value of (H^+) to evaluate both microscopic rate constants.

The termolecular forward reaction conveys no chemical feeling about the intimate molecular details of this reaction, and the proposed mechanism is not conceptually very satisfactory. A mechanism has been proposed in which L is protonated in the first elementary reaction, followed by reaction with solvent in the second elementary reaction. In this mechanism, it is possible for (H^+) to vary with time unless (H^+) is buffered. Set up the chemical and mathematical models for such a mechanism, and see if it is possible in principle to distinguish between the one-reaction and the two-reaction mechanisms on the basis of the dependence of one of the m's on the value of (H^+). In practical experimental terms, can you design an experiment that will distinguish between the two mechanisms? Note that in most of the research papers dealing with this reaction, the rate constants evaluated do not pertain to the two-reaction mechanism that is cited in the same paper as probable.

Search the literature for experimental details concerning the base catalysis of this reaction. The usual starting place for a search of the chemical literature is the weekly publication *Chemical Abstracts* (Columbus, O.: Chemical Abstracts Service, American Chemical Society) which gives a short summary (an abstract) of each chemical research paper which is published in a scholarly journal anywhere in the world. These abstracts are indexed semi-annually by author, by subject, and by chemical formula. Extensive collective indexes are issued every five years. Convenient access to the current chemical literature is via the biweekly *Chemical Titles* (Columbus, O.: Chemical Abstracts Service, American Chemical Society) which publishes English translations of titles of current chemical research articles selected from about 700 journals, indexing these by author and by keywords-in-context.

Pressure-Jump Method: Hydration of Pyruvic Acid in Aqueous Solution. The numerical value of the equilibrium constant for the reaction

$$CH_3-\overset{\overset{O}{\|}}{C}-\overset{\overset{O}{/\!/}}{C}\diagdown_{OH} + H_2O \rightleftharpoons CH_3-\overset{\overset{OH}{|}}{\underset{|}{C}}-\overset{\overset{O}{/\!/}}{\underset{OH}{C}}\diagdown_{OH}$$

is a function of pressure, and therefore the position of equilibrium will change when the pressure is changed. A pressure-jump apparatus for the teaching laboratory has been described [E. M. Eyring, T. W. Keller, and R. P. Jensen, "A Pressure Jump Rate Study of a Bimolecular Reaction," in W. C. Oelke, *Laboratory Physical Chemistry*, New York: Van Nostrand Reinhold Company, 1969] for rapidly changing the position of this equilibrium by a pressure change of about 800 psi within 1 msec. The rate of approach to the new equilibrium is followed by conductance measurement, displayed on an oscilloscope as a trace of conductance versus time and photographed. A single relaxation is observed; the quantity τ used by these authors is equal to $1/m$.

Additional information that has proved useful in constructing this apparatus and in interpreting the results is found in H. Strehlow and H. Wendt, *Inorg. Chem.*, **2**, 6 (1963); H. Hoffmann, J. Stuehr, and E. Yeager, "Studies of Relaxation Effects in Electrolytic Solutions with the Pressure-Step Method," in B. E. Conway and R. G. Barradas, eds., *Chemical Physics of Ionic Solutions*, New York: John Wiley & Sons, Inc., 1966; and M. T. Takahashi, *Pressure-Jump Kinetic Studies on Bovine Plasma Albumin*, University of Wisconsin (Madison) Ph.D. Thesis, 1964; available from University Microfilms, Ann Arbor, Mich., Order No. 64–3944.

A mechanism for the hydration of pyruvic acid is proposed by Eyring, Keller, and Jensen, predicting a dependence of the macroscopic pseudo-first-order rate constant on pH. It would be interesting to perform the experiment at various values of pH and see if the observed values of m_1 do, in fact, vary as predicted. Can you discover what assumptions were made in order to arrive at Equation (2) in the Eyring, Keller, and Jensen chapter?

Can you find out about the hydration of other related compounds? Is there evidence, clear and direct, to support the structure written above for the hydrated acid? What is really known about the kinetics and mechanisms of such reactions? Try to find some answers in the chemical literature. In one relevant article, for example, the pH-variation of the concentrations of species involved in hydration–dehydration reactions of carbonyl compounds in water is analyzed by D. Barnes and P. Zuman [*Talanta*, **16**, 975 (1969)]. For references to additional articles related to these chemical systems, try the latest indexes of *Chemical Abstracts*.

Problems

These problems are arranged in groups in order of increasing difficulty. Answers are found in Appendix VII.

Derive equations relating the macroscopic m's and the microscopic k's for the following mechanisms:

8.1.
$$A + B \underset{k_{-1}}{\overset{k_1}{\rightleftharpoons}} C + D$$

Approximation: constant concentration of A and C.

8.2.
$$A + B^* \underset{k_{-1}}{\overset{k_1}{\rightleftharpoons}} A^* + B$$

Approximation: equilibrium with respect to nonisotopic exchange, with labeled compounds introduced in very low concentration.

8.3.
$$2A \underset{k_{-1}}{\overset{k_1}{\rightleftharpoons}} A_2$$

Approximation: near equilibrium.

8.4.
$$A + B \underset{k_{-1}}{\overset{k_1}{\rightleftharpoons}} C + D$$

Approximation: near equilibrium.

8.5.
$$A + B + C \underset{k_{-1}}{\overset{k_1}{\rightleftharpoons}} D$$

Approximation: near equilibrium.

8.6.
$$A + B \underset{k_{-1}}{\overset{k_1}{\rightleftharpoons}} C$$

$$A + B \underset{k_{-2}}{\overset{k_2}{\rightleftharpoons}} D$$

Approximation: constant concentration of A.

8.7.
$$A + B \underset{k_{-1}}{\overset{k_1}{\rightleftharpoons}} C$$

$$A + B \underset{k_{-2}}{\overset{k_2}{\rightleftharpoons}} D$$

Approximation: near equilibrium.

8.8.
$$A + B \underset{k_{-1}}{\overset{k_1}{\rightleftharpoons}} C$$

$$C \underset{k_{-2}}{\overset{k_2}{\rightleftharpoons}} D$$

Approximation: near equilibrium.

8.9.
$$A + B \underset{k_{-1}}{\overset{k_1}{\rightleftharpoons}} X_1$$

$$X_1 \underset{k_{-2}}{\overset{k_2}{\rightleftharpoons}} X_2$$

$$X_2 \underset{k_{-3}}{\overset{k_3}{\rightleftharpoons}} X_3$$

Approximation: constant concentration of A.

This use of subscripts is commonly used kinetics notation, permitting straightforward labeling of an arbitrarily large number of intermediates. Because the notation of X_2 is identical with the familiar notation for a dimer containing two molecules of X, confusion could result. In the context of the mechanism in this problem, however, the species X_1, X_2, and X_3 differ only in the internal arrangement of atoms within the molecule.

8.10.

$$B \underset{k_{-1}}{\overset{k_1}{\rightleftharpoons}} X_1$$

$$A + X_1 \underset{k_{-2}}{\overset{k_2}{\rightleftharpoons}} X_2$$

$$X_2 \underset{k_{-3}}{\overset{k_3}{\rightleftharpoons}} X_3$$

Approximation: constant concentration of A.

8.11.

$$A + B \underset{k_{-1}}{\overset{k_1}{\rightleftharpoons}} X_1$$

$$A + X_1 \underset{k_{-2}}{\overset{k_2}{\rightleftharpoons}} X_2$$

$$X_2 \underset{k_{-3}}{\overset{k_3}{\rightleftharpoons}} X_3$$

Approximation: constant concentration of A.

8.12.

$$A + B \underset{k_{-1}}{\overset{k_1}{\rightleftharpoons}} X_1$$

$$C + X_1 \underset{k_{-2}}{\overset{k_2}{\rightleftharpoons}} X_2$$

$$X_2 \underset{k_{-3}}{\overset{k_3}{\rightleftharpoons}} X_3 + D$$

Approximation: constant concentration of each of the species A, C, and D.

BIBLIOGRAPHICAL NOTE

The constant-concentration approximation and the near-equilibrium approximation are often used in interpreting rate data for very fast reactions. A thorough treatment of theory is given in G. H. Czerlinski, *Chem-*

ical Relaxation, New York: Marcel Dekker, Inc., 1966. Much information about both theory and experimental techniques is found in E. F. Caldin, *Fast Reactions in Solution*, Oxford: Blackwell Scientific Publications, 1964. A wide-ranging survey of the field is contained in a collection of articles in S. L. Friess, E. S. Lewis, and A. Weissberger, eds., *Investigation of Rates and Mechanisms of Reactions*, part II, New York: John Wiley & Sons, Inc. (Interscience Division), 2nd ed., 1963, chaps. XIV through XXII. E. M. Eyring has written a lucid survey of fast reactions in solution (*Survey of Progress in Chemistry*, vol. 2, New York: Academic Press, Inc., 1964, p. 57). Experimental methods developed for rapid mixing of reactants in fast reactions are described in detail in symposium papers in B. Chance, Q. H. Gibson, R. Eisenhardt, and K. K. Lonberg-Holm, eds., *Rapid Mixing and Sampling Techniques in Biochemistry*, New York: Academic Press, Inc., 1964.

Noting that design of equipment is an integral part of research in chemical kinetics, K. Kustin has edited a volume in the *Methods in Enzymology* series (vol. 16, *Fast Reactions*, New York: Academic Press, Inc., 1969) devoted to construction details of fast-reaction instrumentation. Two of the articles reviewing rapid-mixing methods are Q. H. Gibson, "Rapid Mixing: Stopped Flow," and H. Gutfreund, "Rapid Mixing: Continuous Flow." Experimental methods utilizing the near-equilibrium approximation are described by M. T. Takahashi and R. A. Alberty, "The Pressure-Jump Method," and by T. C. French and G. G. Hammes, "The Temperature-Jump Method."

Application of the near-equilibrium approximation to a mechanism with many sequential elementary reactions is straightforward, but the process of obtaining relationships between the m's and the k's is laborious. G. G. Hammes and P. R. Schimmel [*J. Phys. Chem.*, **70**, 2319 (1966)] point out simplifications that result when no more than two elementary reactions, which equilibrate at comparable rates, are coupled via rapid reactions.

The microscopic rate constants for Reaction (8.51) will be essentially equal to the corresponding k's for Reaction (8.52) if the ratio of masses of the two isotopes involved is close to unity. Significant variation in k values often occurs with hydrogen isotopes, and a great deal of chemical information about the intimate reaction mechanism can be deduced from the quantitative determination of this isotope rate effect. Smaller but measurable isotope effects have been observed and interpreted with the heavier elements, especially carbon, nitrogen, and oxygen. A good overall presentation of kinetic isotope effects is L. Melander, *Isotope Effects on Reaction Rates*, New York: The Ronald Press Company, 1960. A more recent review on kinetic isotope effects is by W. H. Saunders (*Survey of Progress in Chemistry*, vol. 3, New York: Academic Press, Inc., 1966, p. 109). Kinetic isotope effects in organic chemistry and biochemistry are reviewed by H. Simon and D. Palm, *Angew. Chem. Intern. Ed. Engl.*, **5**, 920 (1966).

The solvated proton, symbolized as H^+, is a reactant in many solution reactions in which important elementary reactions are slow compared to proton transfer. It is then convenient to study the slower relaxations under conditions in which the concentration of H^+ is buffered at a constant value, because the constant-concentration approximation can be applied. However, the protonic reactions themselves are of great interest, including some of the fastest known bimolecular elementary reactions. Tabulations of microscopic rate constants for many protonic reactions appear in M. Eigen and L. De Maeyer, "Relaxation Methods," in S. L. Friess, E. S. Lewis, and A. Weissberger, *Investigation of Rates and Mechanisms of Reactions*, part II, New York: John Wiley & Sons, Inc. (Interscience Division), 2nd ed., 1963, pp. 1034–1040, and also in M. Eigen, W. Kruse, G. Maass, and L. De Maeyer, "Rate Constants of Protolytic Reactions in Aqueous Solution," in G. Porter, ed., *Progress in Reaction Kinetics*, vol. 2, New York: The Macmillan Company, 1964. Most of these rate constants were evaluated by use of the near-equilibrium approximation.

9

Enzyme Catalysis by α-Chymotrypsin:

A Case Study

The protein α-chymotrypsin[1] catalyzes a great many hydrolysis reactions, including the cleavage of the peptide bonds which link together amino acids to form a protein molecule. This enzyme also catalyzes the cleavage of many simple amides and esters. We shall examine the kinetics and mechanism of the hydrolysis of the ester, p-nitrophenyl trimethylacetate,[2]

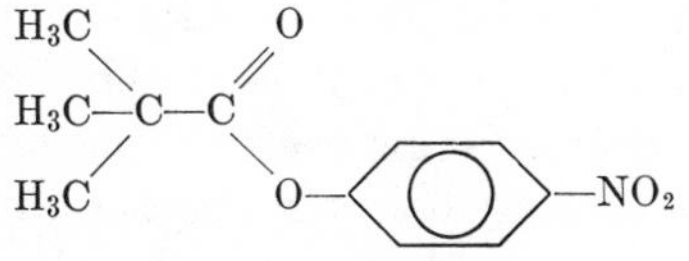

S

to yield p-nitrophenol,

P_1

[1] A series of research articles concerning α-chymotrypsin catalysis is M. L. Bender et al., *J. Amer. Chem. Soc.*, **84**, 2540, 2550, 2562, 2570, 2577, 2582 (1962). This case study follows many of the experimental details of a student experiment described by M. L. Bender, F. J. Kézdy, and F. C. Wedler [*J. Chem. Educ.*, **44**, 84 (1967)].

[2] The use of p-nitrophenyl trimethylacetate and several other nitrophenol esters as substrates for chymotrypsin reactions is reported by C. E. McDonald and A. K. Balls, *J. Biol. Chem.*, **227**, 727 (1957).

and trimethylacetic acid,

$$
\begin{array}{c}
H_3C \\
\quad\diagdown \\
H_3C\!-\!C\!-\!C \\
\quad\diagup \qquad \diagdown \\
H_3C \qquad\qquad OH
\end{array}
$$

P$_2$

The product p-nitrophenol has a substantially different optical absorption spectrum than does the substrate,[3] and the reaction can be followed spectrophotometrically.

Preliminary Experiment.   Solutions were prepared as follows:

THAM–Acetate Buffer.   A 0.01 molar solution of tris(hydroxymethyl)-aminoethane (THAM) in water was prepared, with pH adjusted to pH 8.50 by careful dropwise addition of acetic acid solution. Accuracy in adjusting the pH is much more critical than accuracy in the value of the molarity of THAM. The reaction was later initiated by adding substrate and enzyme to this buffer solution.

Acetic Acid–Acetate Buffer.   A 0.1 molar solution of acetic acid in water was prepared, with pH adjusted to pH 4.6 by careful dropwise addition of sodium hydroxide solution. The enzyme was stored in this buffer solution.

Enzyme Stock Solution.   A 0.0020 molar solution of enzyme was prepared by dissolving 50 mg of three-times crystallized bovine pancreas α-chymotrypsin in 1.0 ml of cold, pH 4.6 acetic acid–acetate buffer. *This solution must be kept refrigerated or in an ice bath.* At pH 4.6 and in the cold, only a few percent decomposition of the enzyme is to be expected during a period as long as 1 week. Caution must be exercised when stirring and pipetting the enzyme solution to prevent foaming and the consequent denaturation and inactivation of the enzyme. Protein solutions behave much like a soap-bubble solution when agitated. Recall the behavior of egg white, a solution of the protein egg albumin in water, when stirred violently with an egg-beater.

Substrate Stock Solution.   A 0.0036 molar solution of the substrate was prepared by dissolving 7.8 mg of p-nitrophenyl trimethylacetate in 10.0 ml of acetonitrile.

The reaction itself was initiated by mixing together 30.00 ml of THAM–acetate buffer, 0.25 ml of substrate stock solution, and 0.25 ml of enzyme stock solution. The reacting solution was then transferred to a

---

[3] The reactant or reactants, in addition to the enzyme, are commonly called the substrate(s) for that enzyme.

cylindrical spectrophotometer cell with an optical path length of 100 mm, and the absorbance of the solution was measured versus time with a Perkin-Elmer Model 350 recording spectrophotometer, using light of wavelength 400 mμ. It was possible to begin recording absorbance within $\frac{1}{2}$ minute after the enzyme solution was added.[4] This plot of absorbance versus time, showing two relaxations, is presented in Figure 9.1.

It turns out that the reaction of ester hydrolysis occurs even without the enzyme present, and so it is necessary to perform a control experiment. For this purpose, 30.00 ml of THAM–acetate buffer, 0.25 ml of substrate stock solution, and 0.25 ml of acetic acid–acetate buffer were mixed and the optical absorbance measured versus time. This plot is also presented in Figure 9.1.

To perform some of the calculations, it is necessary to know the molar absorptivity of the product, p-nitrophenol, in the reaction medium employed for the kinetic run. If at equilibrium there is very little unhydrolyzed ester remaining in solution, the evaluation of the absorptivity of p-nitrophenol involves simply the measurement of the absorbance of the solution

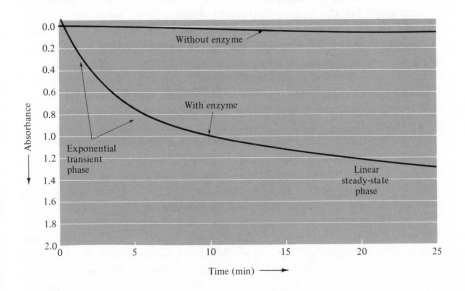

FIGURE 9.1. Absorbance–Time curve showing a transient phase.

<hr/>

[4] Bender, Kézdy, and Wedler [*J. Chem. Educ.*, **44**, 84 (1967)] suggest an alternative method of rapid mixing, employing a much smaller volume of solution. Using a rectangular spectrophotometer cell with a volume of just over 3 ml, the chemist first pours in buffer and substrate solutions. The enzyme solution, just 50 μl in volume, is introduced on the flattened tip of a glass stirring rod. This rod is then used to stir the solution.

used for the control experiment after that solution has reached chemical equilibrium. It is assumed, correctly, that the species trimethylacetate makes only an insignificant contribution to the absorbance at 400 mμ. However, it takes at least 2 weeks for chemical equilibrium to be achieved at room temperature. There is an alternative procedure which takes much less time. In very basic solution, the hydrolysis reaction is much faster. When substrate stock solution is pipetted into sodium hydroxide solution (about 0.1 molar), the color almost immediately becomes yellow. The pH can then be adjusted with acetic acid to the pH value of the reaction mixture, the absorbance measured, and the absorptivity of p-nitrophenol calculated. A third method is even more direct. The compound p-nitrophenol can be dissolved in THAM–acetate buffer, the pH readjusted, and the absorbance determined. No matter whether one of these three procedures is used, or another method tried, it is essential that careful quantitative technique be observed so that a reliable value of the absorptivity can be obtained.

Speculation about Mechanism.   If the absorbance–time curve presented in Figure 9.1 for the enzyme–substrate reaction is to be interpreted in chemical terms, it is necessary to assume a chemical reaction mechanism. The existence of two relaxations requires, as a minimum, that there be two elementary reactions in the mechanism. To keep from having a termolecular reverse process in the mechanism, it, in fact, seems reasonable to include three elementary reactions. The discussion on pages 175 to 178 leads one to anticipate an absorbance–time curve described by an equation of the form (10.133). If the observed curve does in fact have the form (10.133), the transient period of the reaction can be analyzed by means of a modified Guggenheim plot, using Equation (10.141). The subsequent linear portion can then be analyzed by considering that part of the curve to be representing a steady-state situation, and by using an initial-rate, steady-state equation to interpret this straight line.

From these experimental data at one value of substrate concentration, it is possible to use Equations (10.134) and (10.135) to calculate certain groups of kinetic constants. It is desirable, then, to perform such additional experiments at other values of $(S)_0$ as are needed for the evaluation of individual microscopic rate constants, if that is possible, or at least for establishing a range within which certain of the k's must fall. As always, the actual rate experiments yield directly only macroscopic constants. A reaction mechanism must be assumed before any microscopic rate constants can be evaluated, and all such microscopic rate constants have meaning only in the context of the assumed mechanism. Some rather complicated mechanisms have been proposed for this enzyme reaction. With more elementary reactions, and thus more microscopic rate constants, it is less likely that a unique set of k's can be obtained from the experimental data, data that necessarily are accompanied by some experimental uncertainties.

BIBLIOGRAPHICAL NOTE

A short and readable discussion of the general forms of enzyme catalysis is H. Gutfreund and J. R. Knowles, "The Foundations of Enzyme Action," in P. N. Campbell and G. D. Greville, eds., *Essays in Biochemistry*, vol. 3, New York: Academic Press, Inc., 1967, p. 25.

Mechanistic studies of chymotrypsin are discussed and critically interpreted by T. C. Bruice and S. J. Benkovic, *Bioorganic Mechanisms*, vol. I, New York: W. A. Benjamin, Inc., 1966, chap. 2 (available in a paperback edition).

The Steady State

The approach to equilibrium by many chemical systems can be divided into two time phases: an initial transient phase and a steady-state phase. During the steady-state phase, the concentrations of some of the chemical species remain essentially time invariant. The mathematical analysis of the steady-state phase is often greatly simplified by making the steady-state assumption that $d(X_i)/dt = 0$ for certain chemical species X_i.

Some fundamental features of the chemical steady state can be investigated by reexamining a mechanism already discussed in Chapter 5. We shall examine the exact integrated equations, then a limiting case of those exact equations, and finally the steady-state integrated equations for the same mechanism. Consider the reaction

$$A = C \tag{10.1}$$

which proceeds via the two unimolecular elementary reactions

$$A \underset{k_{-1}}{\overset{k_1}{\rightleftharpoons}} B \tag{10.2}$$

$$B \underset{k_{-2}}{\overset{k_2}{\rightleftharpoons}} C \tag{10.3}$$

The three associated differential rate equations are

$$\frac{d(A)}{dt} = -k_1(A) + k_{-1}(B) \tag{10.4}$$

$$\frac{d(B)}{dt} = k_1(A) - k_{-1}(B) - k_2(B) + k_{-2}(C) \tag{10.5}$$

$$\frac{d(C)}{dt} = k_2(B) - k_{-2}(C) \tag{10.6}$$

It was found in Chapter 5 that, for this mechanism, any linear function X of the concentrations (A), (B), and (C) depends on time according to an equation of the form

$$X = x_0 + x_1 e^{-m_1 t} + x_2 e^{-m_2 t} \tag{10.7}$$

where

$$m_1 + m_2 = k_1 + k_{-1} + k_2 + k_{-2} \tag{10.8}$$

$$m_1 m_2 = k_{-1} k_{-2} + k_1 k_2 + k_1 k_{-2} \tag{10.9}$$

There is a limiting case which is of special interest in a consideration of the chemical steady state. Assume the inequality

$$k_{-1} \text{ and } k_2 \text{ individually} \gg k_1 \text{ and } k_{-2} \tag{10.10}$$

For this limiting case, Equations (10.8) and (10.9) become

$$m_1 + m_2 \simeq k_{-1} + k_2 \tag{10.11}$$

$$m_1 m_2 \simeq k_{-1} k_{-2} + k_1 k_2 \tag{10.12}$$

Because of (10.10), the two macroscopic rate constants must have quite different numerical values. We shall label the larger rate constant m_1, and so $m_1 \gg m_2$. Finally, then, we can write

$$m_1 \simeq k_{-1} + k_2 \tag{10.13}$$

$$\frac{m_1 m_2}{m_1 + m_2} \simeq m_2 \simeq \frac{k_{-1} k_{-2} + k_1 k_2}{k_{-1} + k_2} \tag{10.14}$$

For comparison, let us now find out what happens within this set of three differential equations when, instead of assuming inequalities among the k's, the steady-state assumption is assumed in the form

$$\frac{d(B)}{dt} = 0 \tag{10.15}$$

Equation (10.5) becomes

$$(B) = \frac{k_1(A) + k_{-2}(C)}{k_{-1} + k_2} \tag{10.16}$$

Introduction of Equation (10.16) into differential equation (10.4) results in

$$\frac{d(A)}{dt} = -k_1(A) + \frac{k_{-1}k_1(A)}{k_{-1} + k_2} + \frac{k_{-1}k_{-2}(C)}{k_{-1} + k_2}$$

$$= \left(\frac{-k_1k_{-1} - k_1k_2 + k_{-1}k_1}{k_{-1} + k_2} \right)(A) + \frac{k_{-1}k_{-2}(C)}{k_{-1} + k_2} \tag{10.17}$$

Equations (10.16) and (10.6) together yield

$$\frac{d(C)}{dt} = \frac{k_1k_2(A)}{k_{-1} + k_2} + \left(\frac{k_2k_{-2} - k_{-1}k_{-2} - k_2k_{-2}}{k_{-1} + k_2} \right)(C) \tag{10.18}$$

Introduction into Equations (10.17) and (10.18) of the trial solutions

$$(A) = ae^{-mt} \tag{10.19}$$

$$(C) = ce^{-mt} \tag{10.20}$$

gives

$$a\left(\frac{k_1k_2}{k_{-1} + k_2} - m \right) - c\left(\frac{k_{-1}k_{-2}}{k_{-1} + k_2} \right) = 0 \tag{10.21}$$

$$-a\left(\frac{k_1k_2}{k_{-1} + k_2} \right) + c\left(\frac{k_{-1}k_{-2}}{k_{-1} + k_2} - m \right) = 0 \tag{10.22}$$

These two simultaneous equations have the roots

$$m_0' = 0 \tag{10.23}$$

$$m_1' = \frac{k_1k_2 + k_{-1}k_{-2}}{k_{-1} + k_2} \tag{10.24}$$

Equation (10.24) is identical with Equation (10.14), provided only that we identify m_2 of the complete solution with the one nonzero root, m_1', of the steady-state solution. If we do this, we can make some generalizations and read some chemistry into the preceding algebra.

This reaction proceeds in two phases. There is a fast relaxation—a transient phase—during which there is only a very small percentage change in the concentration of A and very little production of C. During this transient phase, however, the value of (B) reaches its steady-state value. When the steady state with respect to (B) is achieved, Equation (10.15) is valid, and then the remainder of the approach to equilibrium can be described by a single relaxation process characterized by m_2 of Equation (10.14) or m_1' of Equation (10.24). The first-order, steady-state approach to equilibrium is a process involving the entire reacting system, requiring the inclusion of both reactions, described quantitatively by all four microscopic rate constants, and not separable into contributions from either (10.2) or (10.3) alone.

During the steady-state phase of the reaction, the molecules of B throughout the solution collectively constitute a special sort of open system in a steady state. There is input to the system as reactant A is transformed into molecules of B. There is output from the system as B is converted into product C. In spite of the relatively large throughput of molecules, the number of molecules of intermediate B in this steady-state system remains nearly constant in time.

The distinction between steady state and equilibrium is important. Establishment of a steady state with respect to (B) still allows an essentially one-way flow of material from A to C during early periods of the steady-state phase. At equilibrium, the number of molecules of C transferred through B to become A is equal to the number of molecules of A transferred through B to become C, during any period of time. We have here represented successively more restrictive levels of constraints upon the system, that is, successively more restrictive levels of time invariance within the reacting system. In general, all concentrations within a reacting system are changing with time. In the steady-state approximation, certain concentrations are assumed to be time invariant. A higher level of invariance occurs when each concentration has a time-invariant value, and this invariance is achieved at chemical equilibrium. In addition to this restriction on concentrations, an additional restriction on the values of the rate constants is required when the principle of microscopic reversibility is invoked at equilibrium. Thus time invariance of each concentration is required for the elementary reactions considered individually and independently.

Use of words such as input, output, throughput, and flow can be very descriptive. Remember, however, that this reaction is taking place in all regions of the solution. No net migration of molecules through the solution is implied by the term "flow." These molecules, in the act of being transformed by a series of chemical changes, create a flow of material through a process, through a chemical reaction.

Enzyme Catalysis

There are a great many reactions of biochemical importance catalyzed by enzymes in which the overall stoichiometric reaction can be formulated as

$$E + S = P + E \tag{10.25}$$

where E is the enzyme, S is the reactant (commonly called the substrate), and P is the product. It is difficult to think about a specific chemical reaction of the type (10.25) without imagining some sort of intermediate complex. It seems reasonable that there might exist several complexes intermediate in structure between the two collision partnerships $E \cdots S$ and $E \cdots P$. To what extent can experimental studies of the rate of an enzyme-catalyzed reaction give information about the existence and nature of enzyme–reactant complexes? A great deal of current research activity is directed toward answering this question. We can gain insight into this research by examining several alternative mechanisms by which (10.25) could take place.

Mechanism with No Intermediate. The simplest possible mechanism for Reaction (10.25) has no intermediate, and it can be formulated as

$$E + S \underset{k_{-1}}{\overset{k_1}{\rightleftharpoons}} P + E \tag{10.26}$$

The two differential rate equations for the mechanism are

$$\frac{d(S)}{dt} = -k_1(E)(S) + k_{-1}(E)(P) \tag{10.27}$$

$$\frac{d(P)}{dt} = k_1(E)(S) - k_{-1}(E)(P) \tag{10.28}$$

Even though E is both a reactant and a product, it never gets converted into anything, and so the quantity (E) has a time-invariant value. To indicate explicitly that (E) always has its initial-time value, $(E)_0$, we write

$$(E) = (E)_0 \tag{10.29}$$

Two trial solutions are assumed:

$$(S) = se^{-mt} \tag{10.30}$$

$$(P) = pe^{-mt} \tag{10.31}$$

Substitution of Equations (10.29), (10.30), and (10.31) into differential equations (10.27) and (10.28) yields, after cancellation of the common exponential factor,

$$s[k_1(E)_0 - m] - p[k_{-1}(E)_0] \qquad = 0 \qquad (10.32)$$

$$-s[k_1(E)_0] \qquad + p[k_{-1}(E)_0 - m] = 0 \qquad (10.33)$$

The two solutions of these simultaneous algebraic equations are

$$m_0 = 0 \qquad (10.34)$$

$$m_1 = (E)_0[k_1 + k_{-1}] \qquad (10.35)$$

and Equations (10.30) and (10.31) become

$$(S) = s_0 + s_1 e^{-m_1 t} \qquad (10.36)$$

$$(P) = p_0 + p_1 e^{-m_1 t} \qquad (10.37)$$

Thus, in the absence of an intermediate, the bimolecular reaction (10.26) proceeds via a first-order relaxation process.

Mechanism with One Intermediate. Another reaction mechanism which would yield the overall stoichiometry of (10.25) is

$$E + S \underset{k_{-1}}{\overset{k_1}{\rightleftharpoons}} X \qquad (10.38)$$

$$X \underset{k_{-2}}{\overset{k_2}{\rightleftharpoons}} P + E \qquad (10.39)$$

The species X is a molecular complex containing the enzyme. The rate constants, as always, have meaning only with respect to this particular reaction mechanism, and there is no necessary correspondence with the rate constants of Mechanism (10.26).

The four differential rate equations are

$$\frac{d(E)}{dt} = -k_1(E)(S) - k_{-2}(E)(P) + [k_2 + k_{-1}](X) \qquad (10.40)$$

$$\frac{d(X)}{dt} = k_1(E)(S) + k_{-2}(E)(P) - [k_2 + k_{-1}](X) \qquad (10.41)$$

$$\frac{d(S)}{dt} = -k_1(E)(S) + k_{-1}(X) \qquad (10.42)$$

$$\frac{d(P)}{dt} = -k_{-2}(E)(P) + k_2(X) \qquad (10.43)$$

The exact solution of these simultaneous nonlinear differential equations has never been obtained in a form valid for arbitrary values of the concentrations and of the rate constants, and in a form valid throughout the entire course of the reaction. If we are to gain insight into the mathematics of this enzymatic reaction, we should restrict our attention at first to some special instances in which the differential equations can be solved. Therefore, we shall examine next the special case in which $k_1 = k_{-2}$, $(S)_0 \gg (E)_0$, a case that results in first-order relaxation kinetics. A steady-state solution for this special case will also be obtained in an integrated form, and the two solutions will be compared.

1. $k_1 = k_{-2}$, $[(S)_0 + (P)_0] \gg (E)_0$. We shall first obtain equations for the time dependence of the concentrations (E) and (X). Equations (10.40) and (10.41) will both be written in terms of k_1, giving

$$\frac{d(E)}{dt} = -k_1[(S) + (P)](E) + [k_2 + k_{-1}](X) \qquad (10.44)$$

$$\frac{d(X)}{dt} = k_1[(S) + (P)](E) - [k_2 + k_{-1}](X) \qquad (10.45)$$

The individual concentrations (S) and (P) undergo marked changes during the reaction, but the sum $[(S) + (X) + (P)]$ is a time-invariant quantity. Since the value of (X) can never exceed the initial value of (E), and since $[(S)_0 + (P)_0]$ is much greater than $(E)_0$, it is valid to make the approximation

$$(S) + (P) \simeq (S) + (X) + (P) \qquad (10.46)$$

and to consider the quantity $[(S) + (P)]$ essentially time invariant. Now assume the trial solutions

$$(E) = \varepsilon e^{-mt} \qquad (10.47)$$

$$(X) = x e^{-mt} \qquad (10.48)$$

Equations (10.47) and (10.48) are substituted into (10.44) and (10.45):

$$\varepsilon\{k_1[(S) + (P)] - m\} - x\{k_2 + k_{-1}\} = 0 \qquad (10.49)$$

$$-\varepsilon\{k_1[(S) + (P)]\} + x\{k_2 + k_{-1} - m\} = 0 \qquad (10.50)$$

The two roots of these simultaneous equations are

$$m_0 = 0 \qquad (10.51)$$

$$m_1 = k_1[(S) + (P)] + k_{-1} + k_2 \qquad (10.52)$$

Equations (10.47) and (10.48) become

$$(E) = \varepsilon_0 + \varepsilon_1 e^{-m_1 t} \tag{10.53}$$

$$(X) = x_0 + x_1 e^{-m_1 t} \tag{10.54}$$

Later the macroscopic rate constant m_1 will be identified as the transient-phase rate constant. Considered in this way, Equations (10.53) and (10.54) describe the relaxation of the concentrations of E and X from the initial values to the steady-state values during the pre-steady-state transient phase.

Integrated equations for the time dependence of (S) and (P) can be obtained,[1] and it is found that both concentrations can be expressed in terms of two first-order relaxation processes. Thus

$$(P) = p_0 + p_1 e^{-m_1 t} + p_2 e^{-m_2 t} \tag{10.55}$$

$$(S) = s_0 + s_1 e^{-m_1 t} + s_2 e^{-m_2 t} \tag{10.56}$$

where m_1 is given by Equation (10.52), and m_2 is

$$m_2 = \frac{[k_1 k_{-1} + k_1 k_2](E)_0}{k_1[(S) + (P)] + k_{-1} + k_2} \tag{10.57}$$

Equations (10.55) and (10.56), valid throughout the reaction, include separate terms for the transient phase and for the steady-state phases. What is the justification for asserting that the macroscopic rate constant m_2 is associated with the steady-state relaxation of the reaction? A straightforward approach to an answer would be to assume the steady-state approximation for species E and X, then to derive equations corresponding to Equations (10.55)–(10.57), and finally to compare the results with the exact equations. This is just what we shall do.

2. E and X assumed in steady state, $k_1 = k_{-2}$. The steady-state assumption is introduced by setting the derivative $d(E)/dt$ equal to zero in Equation (10.44), or by setting the derivative $d(X)/dt$ equal to zero in Equation (10.45). These alternative procedures are equivalent. It will also be convenient to write a conservation equation

$$(E)_0 = (E) + (X) \tag{10.58}$$

---

[1] W. G. Miller and R. A. Alberty, *J. Amer. Chem. Soc.*, **80**, 5146 (1958). In this paper the exact solutions to the rate equations for Mechanism (10.38)–(10.39) are derived for the case $k_1 = k_{-2}$. The steady-state approximation is shown to be a good approximation if $(S)_0 \gg (E)_0$ or if $[(E)_0 + (S)_0] \ll [k_{-1} + k_2]/k_1$. A perturbation solution is developed for the case $k_1 \neq k_{-2}$, and the applicability of the steady-state approximation for this case is discussed.

The steady-state condition applied to either Equation (10.44) or Equation (10.45) yields, with $k_1 = k_{-2}$,

$$(E) = \frac{[k_2 + k_{-1}](X)}{k_1[(S) + (P)]} \tag{10.59}$$

When Equations (10.58) and (10.59) are combined, the concentration (E) is eliminated, and one gets

$$(X) = \frac{(E)_0 k_1[(S) + (P)]}{k_1[(S) + (P)] + k_2 + k_{-1}} \tag{10.60}$$

The conservation equation (10.58) is also used to eliminate (E) from differential equations (10.42) and (10.43), with the results

$$\frac{d(S)}{dt} = -k_1(E)_0(S) + k_1(X)(S) + k_{-1}(X) \tag{10.61}$$

$$\frac{d(P)}{dt} = -k_1(E)_0(P) + k_1(X)(P) + k_2(X) \tag{10.62}$$

Each differential equation has a second-order term. Linearization of the differential equations is now achieved by using Equation (10.60) to eliminate the concentration (X) from both (10.61) and (10.62). The desired differential equations are

$$\frac{d(S)}{dt} = \frac{-k_1 k_2(E)_0(S)}{k_1[(S) + (P)] + k_2 + k_{-1}} + \frac{k_1 k_{-1}(E)_0(P)}{k_1[(S) + (P)] + k_2 + k_{-1}} \tag{10.63}$$

$$\frac{d(P)}{dt} = \frac{k_1 k_2(E)_0(S)}{k_1[(S) + (P)] + k_2 + k_{-1}} + \frac{-k_1 k_{-1}(E)_0(P)}{k_1[(S) + (P)] + k_2 + k_{-1}} \tag{10.64}$$

The quantity $[(S) + (P)]$ is essentially time invariant, and thus each term contains just one time-dependent concentration factor. The following trial solutions are then assumed:

$$(S) = se^{-mt} \tag{10.65}$$

$$(P) = pe^{-mt} \tag{10.66}$$

Introduction of the trial solutions into Equations (10.63) and (10.64) produces a pair of simultaneous algebraic equations. There are two roots,

$$m_0' = 0 \tag{10.67}$$

$$m_1' = \frac{[k_1 k_{-1} + k_1 k_2](E)_0}{k_1[(S) + (P)] + k_2 + k_{-1}} \tag{10.68}$$

Equations (10.68) and (10.57) are identical, except for the labeling of the macroscopic rate constant. This identity is our justification for saying that the exponential terms

$$p_2 e^{-m_2 t} \quad \text{and} \quad s_2 e^{-m_2 t}$$

in Equations (10.55) and (10.56) are associated with a first-order steady-state relaxation process. This identity is also the justification for saying that it is valid to assume the steady-state approximation for this special case, because the derivation with the steady-state assumption gives results equivalent to the results from the exact derivation for the time period after the transient phase.

Initial-Rate Equations. We shall now remove the restriction that two rate constants are equal, and examine the steady-state rate equations for Reaction (10.25), considering mechanisms with no intermediate, one intermediate, two intermediates, and finally any finite number of intermediates. In each case, the results will be presented in a form suitable for initial-rate experiments.

No Intermediate. The steady state with respect to enzyme E is implicit in Mechanism (10.26). The assumption that $d(E)/dt$ equals zero is a direct consequence of the fact that $(E)_0 = (E)$. Combination of Equations (10.27) and (10.29) yields

$$\frac{d(S)}{dt} = -(E)_0[k_1(S) - k_{-1}(P)] \tag{10.69}$$

If the reaction begins without product P present, the conditions at the initial time are

$$(S) = (S)_0 \qquad (P) = 0 \tag{10.70}$$

The initial rate, the rate of the reaction at time zero, is then

$$\left(\frac{d(S)}{dt}\right)_0 = -k_1(E)_0(S)_0 \tag{10.71}$$

If the reaction had been initiated from the opposite direction, under initial conditions of

$$(S) = 0 \qquad (P) = (P)_0 \qquad\qquad (10.72)$$

then the initial rate would have been given by

$$\left(\frac{d(S)}{dt}\right)_0 = k_{-1}(E)_0(P)_0 \qquad\qquad (10.73)$$

One Intermediate. Application of the steady-state approximation to intermediate X in Mechanism (10.38)–(10.39) requires that the derivative $d(X)/dt$ in Equation (10.41) be set equal to zero. This is equivalent to assuming a steady state with respect to free enzyme E, and setting the derivative $d(E)/dt$ in Equation (10.40) equal to zero. The result in either case is

$$(X) = \frac{k_1(E)(S) + k_{-2}(E)(P)}{k_2 + k_{-1}} \qquad\qquad (10.74)$$

Combination of (10.74) with conservation equation (10.58) permits elimination of either (X) or (E), giving

$$(E) = (E)_0\left(\frac{k_2 + k_{-1}}{k_2 + k_{-1} + k_1(S) + k_{-2}(P)}\right) \qquad\qquad (10.75)$$

$$(X) = (E)_0\left(\frac{k_1(S) + k_{-2}(P)}{k_2 + k_{-1} + k_1(S) + k_{-2}(P)}\right) \qquad\qquad (10.76)$$

Equations (10.75) and (10.76) can then be used to eliminate the two concentrations (E) and (X) from Equation (10.42), leaving

$$\frac{d(S)}{dt} = \left(\frac{k_{-1}k_{-2}(P) - k_1k_2(S)}{k_2 + k_{-1} + k_1(S) + k_{-2}(P)}\right)(E)_0 \qquad\qquad (10.77)$$

It is customary to group certain of the rate constants in Equation (10.77) by defining four new quantities,

$$V_S = k_2(E)_0 \qquad\qquad (10.78)$$

$$V_P = k_{-1}(E)_0 \qquad\qquad (10.79)$$

$$K_S = \frac{k_{-1} + k_2}{k_1} \qquad\qquad (10.80)$$

$$K_P = \frac{k_{-1} + k_2}{k_{-2}} \qquad\qquad (10.81)$$

Written in terms of V_S, V_P, K_S, and K_P, Equation (10.77) becomes

$$\frac{d(S)}{dt} = \frac{\left(\dfrac{V_P}{K_P}\right)(P) - \left(\dfrac{V_S}{K_S}\right)(S)}{1 + \dfrac{(S)}{K_S} + \dfrac{(P)}{K_P}} \tag{10.82}$$

This equation is instantaneously valid throughout the entire steady-state of the reaction, but it is not very useful for making a quantitative study of the rate of the reaction. The equation is easily cast into a limiting form for initial-rate measurements. For instance, if at initial time $(S) = (S)_0$ and $(P) = 0$, then the initial rate, $(d(S)/dt)_0$, is given by

$$\left(\frac{d(S)}{dt}\right)_0 = - \frac{\left(\dfrac{V_S}{K_S}\right)(S)_0}{1 + \dfrac{(S)_0}{K_S}} = - \frac{V_S(S)_0}{K_S + (S)_0} \tag{10.83}$$

If it is possible to observe the rate of the reaction [the slope of an (S) versus t plot, which is the value of the derivative, $(d(S)/dt)_0$] under conditions such that during the time interval of measurement,

$$(S) \simeq (S)_0$$

$$\left(\frac{V_P}{K_P}\right)(P) \ll \left(\frac{V_S}{K_S}\right)(S)$$

$$\frac{(P)}{K_P} \ll \frac{(S)}{K_S}$$

$$\frac{d(X)}{dt} = 0$$

then a series of experiments, a series of kinetic runs, can be planned for the determination of the numerical values of V_S and K_S. The numbers from experiments will be ordered pairs of numbers $\{(d(S)/dt)_0, (S)_0\}$. A convenient graphical method for using these experimental data to evaluate both V_S and K_S employs Equation (10.83) in the reciprocal form,

$$\frac{1}{\left(\dfrac{d(S)}{dt}\right)_0} = - \frac{K_S + (S)_0}{V_S(S)_0} = - \left(\frac{K_S}{V_S}\right)\left(\frac{1}{(S)_0}\right) - \frac{1}{V_S} \tag{10.84}$$

A plot of the reciprocal of the initial rate versus $[1/(S)_0]$ should yield a straight line with slope equal to $-K_S/V_S$ and intercept equal to $-[1/V_S]$.

If it is possible to determine the initial rate of the reaction under steady-state conditions, starting with just P and no S in the reaction mixture, then the analogous equations permit numerical evaluation of K_P and V_P. Knowledge of the values of the four macroscopic kinetic constants permits use of Equations $(10.78) - (10.81)$ to evaluate the four microscopic rate constants of the mechanism, if the numerical value of $(E)_0$ is known with a reasonable degree of confidence. The value of $(E)_0$ is often uncertain, because of substantial difficulties in isolating and purifying some enzymes, because even crystalline enzyme preparations may contain contaminating nonenzyme proteins or inactivated enzyme, because of the instability of many enzyme preparations and enzyme solutions, and because of some ambiguity in the definition of (E) when the enzyme molecule is capable of simultaneously reacting with more than one molecule of substrate.

Two Intermediates.   The inclusion of two isomeric intermediates[2] X_1 and X_2 in the enzymatic mechanism results in a three-reaction mechanism,

$$E + S \underset{k_{-1}}{\overset{k_1}{\rightleftharpoons}} X_1 \tag{10.85}$$

$$X_1 \underset{k_{-2}}{\overset{k_2}{\rightleftharpoons}} X_2 \tag{10.86}$$

$$X_2 \underset{k_{-3}}{\overset{k_3}{\rightleftharpoons}} E + P \tag{10.87}$$

The associated differential rate equations are

$$\frac{d(S)}{dt} = -k_1(E)(S) + k_{-1}(X_1) \tag{10.88}$$

$$\frac{d(E)}{dt} = -k_1(E)(S) + k_{-1}(X_1) + k_3(X_2) - k_{-3}(E)(P) \tag{10.89}$$

$$\frac{d(X_1)}{dt} = k_1(E)(S) - k_{-1}(X_1) - k_2(X_1) + k_{-2}(X_2) \tag{10.90}$$

$$\frac{d(X_2)}{dt} = k_2(X_1) - k_{-2}(X_2) - k_3(X_2) + k_{-3}(E)(P) \tag{10.91}$$

$$\frac{d(P)}{dt} = k_3(X_2) - k_{-3}(E)(P) \tag{10.92}$$

[2] This use of subscripts is commonly employed kinetics notation, permitting straightforward labeling of an arbitrarily large number of intermediates. Because the notation for X_2 is identical with the familiar notation for a dimer containing two molecules of X, confusion could result. In the context of Equations (10.85)–(10.87), however, the species X_1 and X_2 differ only in the internal arrangement of atoms.

The conservation equation, summing the concentrations of all species containing the enzyme, is

$$(E)_0 = (E) + (X_1) + (X_2) \tag{10.93}$$

The steady-state approximation is introduced by setting the derivatives in Equations (10.89), (10.90), and (10.91) equal to zero. The resulting algebraic equations can then be combined to give

$$(X_1)\{k_{-1}k_{-2} + k_{-1}k_3 + k_2k_3\}$$
$$= (E)\{[k_1k_{-2} + k_1k_3](S) + k_{-2}k_{-3}(P)\} \tag{10.94}$$

$$(X_2)\{k_{-1}k_{-2} + k_{-1}k_3 + k_2k_3\}$$
$$= (E)\{k_1k_2(S) + [k_{-1}k_{-3} + k_2k_{-3}](P)\} \tag{10.95}$$

$$(X_2)\{[k_1k_{-2} + k_1k_3](S) + k_{-2}k_{-3}(P)\}$$
$$= (X_1)\{k_1k_2(S) + [k_{-3}k_{-1} + k_2k_{-3}](P)\} \tag{10.96}$$

Equations (10.93)–(10.96) can be combined to give (E) and (X_1) each as a function of $(E)_0$, (S), (P), and the rate constants. Substitution into Equation (10.88) then gives

$$\frac{d(S)}{dt} = \frac{(E)_0[k_{-1}k_{-2}k_{-3}(P) - k_1k_2k_3(S)]}{k_{-1}k_{-2} + k_{-1}k_3 + k_2k_3 + k_1(S)[k_{-2} + k_3 + k_2] + k_{-3}(P)[k_{-1} + k_{-2} + k_2]} \tag{10.97}$$

The remarkable fact is that Equation (10.97) can be written in exactly the same form as Equation (10.82), the steady-state differential rate equation for a single-intermediate mechanism. The required change of variables to make the two equations look the same is

$$K_P = \frac{k_{-1}k_{-2} + k_{-1}k_3 + k_2k_3}{k_{-3}[k_{-1} + k_{-2} + k_2]} \tag{10.98}$$

$$K_S = \frac{k_{-1}k_{-2} + k_{-1}k_3 + k_2k_3}{k_1[k_2 + k_{-2} + k_3]} \tag{10.99}$$

$$V_P = \frac{k_{-1}k_{-2}(E)_0}{k_{-1} + k_{-2} + k_2} \tag{10.100}$$

$$V_S = \frac{k_2k_3(E)_0}{k_2 + k_{-2} + k_3} \tag{10.101}$$

The two different sets of macroscopic steady-state constants K_S, K_P, V_S, and V_P have different meanings in terms of the microscopic rate constants of the two different reaction mechanisms. But since either mechanism results in Equation (10.82), it has turned out that steady-state kinetic measurements cannot distinguish between alternative mechanisms with one or two intermediates. Equation (10.82) is, in fact, much more general than has been indicated. This equation results for any mechanism in which the first step is like (10.85), the last step is like (10.87), and in which some finite number of sequential unimolecular isomerizations like (10.86) lead from the first step to the last.[3] The incapacity of steady-state experiments to give information in detail about this sequence of intermediates is a direct consequence of the fact that, throughout the time of observation, all these intermediate species remain in their steady-state concentrations. Steady-state rate measurements give the rate of throughput in an essentially time-invariant system of reaction intermediates, a system that consumes substrate S and yields product P. The fine structure of this system, described in terms of the chemical identity of the intermediates, of the intimate mechanism of interconversion of intermediates, and of the numerical values of the individual microscopic rate constants, is accessible only through kinetic measurements of systems in which the intermediates of interest have not yet reached their steady-state concentrations.

We have just seen that by including additional intermediates we can write many alternative mechanisms, some quite complex, which are consistent with (but certainly not required by) the limited amount of information contained in experimental steady-state rate data. The temptation is sometimes great to devise an elaborate, richly detailed mechanistic scheme that is indistinguishable in terms of its verifiable predictions from the minimal mechanisms required by the kinetic observations. Some common sense should be brought to bear on any conceptual model that seems to contain appreciably more information than that provided by observation of the phenomenon itself. Samuel Clemens gives an example.

> In the space of one hundred and seventy-six years the Lower Mississippi has shortened itself two hundred and forty-two miles. That is an average of a trifle over one mile and a third per year. Therefore, any calm person, who is not blind or idiotic, can see that in the Old Oölitic Silurian Period, just a million years ago next November, the Lower Mississippi River was upwards of one million three hundred thousand miles long, and stuck out over the Gulf of Mexico like a fishing-rod. And by the same token any person can see that seven hundred and forty-two years from now the Lower Mississippi will be only a mile and three quarters long, and Cairo and New Orleans will have joined their streets together,

[3] L. Peller and R. A. Alberty, *J. Amer. Chem. Soc.*, **81**, 5907 (1959).

and be plodding comfortably along under a single mayor and a mutual board of aldermen. There is something fascinating about science. One gets such wholesale returns of conjecture out of such a trifling investment of fact.[4]

Transient-Phase Kinetics.   Steady-state kinetic studies alone are not sufficient to reveal the details of an enzymatic mechanism. Rates of reactions involving the enzyme α-chymotrypsin have been studied under steady-state conditions and also during the time before the steady state is established. This pre-steady-state period, from the beginning of the reaction until the steady-state concentrations have been established, is often called the transient phase of the reaction. The additional information obtained for α-chymotrypsin reactions during the transient phase supplements the steady-state rate data for the same reaction, and it has thus been possible to evaluate some of the microscopic rate constants for a proposed mechanism.

A reaction mechanism that has been proposed by many of those chemists[5] who have studied α-chymotrypsin catalysis is

$$E + S \underset{k_{-1}}{\overset{k_1}{\rightleftharpoons}} ES' \tag{10.102}$$

$$ES' \underset{k_{-2}}{\overset{k_2}{\rightleftharpoons}} ES'' + P' \tag{10.103}$$

$$ES'' \underset{k_{-3}}{\overset{k_3}{\rightleftharpoons}} E + P'' \tag{10.104}$$

The differential equations associated with the mechanism are

$$\frac{d(E)}{dt} = -k_1(E)(S) + k_{-1}(ES') + k_3(ES'') - k_{-3}(E)(P'') \tag{10.105}$$

$$\frac{d(ES')}{dt} = k_1(E)(S) - k_{-1}(ES') - k_2(ES') + k_{-2}(ES'')(P') \tag{10.106}$$

$$\frac{d(ES'')}{dt} = k_2(ES') - k_{-2}(ES'')(P') - k_3(ES'') + k_{-3}(E)(P'') \tag{10.107}$$

$$\frac{d(S)}{dt} = -k_1(E)(S) + k_{-1}(ES') \tag{10.108}$$

[4] M. Twain, *The Atlantic Monthly*, **36**, 193 (1875).
[5] See, for instance, the series of articles by M. L. Bender and his research associates, *J. Amer. Chem. Soc.*, **84**, 2540, 2550, 2562, 2570, 2577, 2582 (1962).

$$\frac{d(P')}{dt} = k_2(ES') - k_{-2}(ES'')(P') \tag{10.109}$$

$$\frac{d(P'')}{dt} = k_3(ES'') - k_{-3}(E)(P'') \tag{10.110}$$

The conservation equation for enzyme-containing species is

$$(E)_0 = (E) + (ES') + (ES'') \tag{10.111}$$

For this set of simultaneous equations, there are no exact solutions which are valid throughout the entire time course of the reaction. It is necessary to make some sort of mathematical approximation to obtain any solution, and therefore any solution that we can write down in algebraic form will be an approximate solution. By making the steady-state assumption, we shall obtain an approximate solution valid for all times after the steady state has been established. By making some very drastic assumptions, we shall be able to obtain an approximate transient-phase solution. However, these assumptions are difficult to justify in quantitative experimental terms.

Steady-State Solution. The assumption is made that all enzyme-containing species have relaxed to their steady-state concentrations. This is the mathematical approximation that $d(E)/dt = 0$, $d(ES')/dt = 0$, and $d(ES'')/dt = 0$. Then Equations (10.105), (10.106), and (10.107) become

$$(E)[k_1(S) + k_{-3}(P'')] = k_{-1}(ES') + k_3(ES'') \tag{10.112}$$

$$(ES')[k_{-1} + k_2] = k_1(E)(S) + k_{-2}(ES'')(P') \tag{10.113}$$

$$(ES'')[k_{-2}(P') + k_3] = k_{-3}(E)(P'') + k_2(ES') \tag{10.114}$$

Combination of Equations (10.113), (10.114), and (10.111) yields

$$(ES') = \frac{(E)_0[k_1k_3(S) + k_1k_{-2}(S)(P') + k_{-2}k_{-3}(P')(P'')]}{\begin{aligned}&k_{-1}k_3 + k_2k_3 + k_1(S)[k_3 + k_2 + k_{-2}(P')]\\&+ k_{-3}(P'')[k_{-1} + k_2 + k_{-2}(P')] + k_{-1}k_{-2}(P')\end{aligned}} \tag{10.115}$$

Combination of Equations (10.113), (10.114), and (10.111) also yields

$$(ES'') = \frac{(E)_0[k_1k_2(S) + k_{-1}k_{-3}(P'') + k_2k_{-3}(P'')]}{\begin{aligned}&k_{-1}k_3 + k_2k_3 + k_1(S)[k_3 + k_2 + k_{-2}(P')]\\&+ k_{-3}(P'')[k_{-1} + k_2 + k_{-2}(P')] + k_{-1}k_{-2}(P')\end{aligned}} \tag{10.116}$$

These two steady-state concentrations of intermediates can be substituted into Equation (10.109), giving

$$\frac{d(\mathrm{P'})}{dt} = \frac{(\mathrm{E})_0[k_1 k_2 k_3(\mathrm{S}) - k_{-1} k_{-2} k_{-3}(\mathrm{P'})(\mathrm{P''})]}{\begin{array}{c} k_{-1} k_3 + k_2 k_3 + k_1(\mathrm{S})[k_3 + k_2 + k_{-2}(\mathrm{P'})] \\ + k_{-3}(\mathrm{P''})[k_{-1} + k_2 + k_{-2}(\mathrm{P'})] + k_{-1} k_{-2}(\mathrm{P'}) \end{array}} \tag{10.117}$$

The same procedure can be used to obtain the derivatives $-d(\mathrm{S})/dt$ and $d(\mathrm{P''})/dt$ as a function of the microscopic rate constants and the three time-dependent concentrations (S), $(\mathrm{P'})$, and $(\mathrm{P''})$. In the steady state, all three derivatives are equal.

If it is possible to measure the rate of formation of $\mathrm{P'}$ *after* the steady state is established, but *before* $(\mathrm{P'})$ and $(\mathrm{P''})$ have become large enough so that terms containing these concentrations contribute in Equation (10.117), and *before* (S) is significantly different from $(\mathrm{S})_0$, then we can write

$$\left(\frac{d(\mathrm{P'})}{dt}\right)_{t=0,\,\text{steady state}} = \frac{(\mathrm{E})_0(\mathrm{S})_0 k_1 k_2 k_3}{(\mathrm{S})_0[k_1 k_2 + k_1 k_3] + k_{-1} k_3 + k_2 k_3} \tag{10.118}$$

A solution for differential equation (10.118) is

$$(\mathrm{P'}) = \left(\frac{(\mathrm{E})_0(\mathrm{S})_0 k_1 k_2 k_3}{(\mathrm{S})_0[k_1 k_2 + k_1 k_3] + k_{-1} k_3 + k_2 k_3}\right) t \tag{10.119}$$

It is clear that the combination of circumstances required to make Equation (10.119) useful may not always result. The equation can never be strictly valid, because both (10.118) and (10.119) require simultaneously that t be equal to zero and that the transient phase has already been completed. However, when applicable, Equation (10.119) suggests a simple treatment of experimental rate data. A plot of $(\mathrm{P'})$ versus t should be a straight line.

Transient-Phase Solution. Following Gutfreund and Sturtevant,[6] we make the approximations that, in the early stages of the reaction

$$k_2(\mathrm{ES'}) \gg k_{-2}(\mathrm{ES''})(\mathrm{P'}) \tag{10.120}$$

$$k_3(\mathrm{ES''}) \gg k_{-3}(\mathrm{E})(\mathrm{P''}) \tag{10.121}$$

$$(\mathrm{S}) = (\mathrm{S})_0 \tag{10.122}$$

Differential equations (10.105), (10.106), (10.107), and (10.109) are

---

[6] H. Gutfreund and J. M. Sturtevant, *Biochem. J.*, **63**, 656 (1956).

linearized by the approximations, becoming

$$\frac{d(E)}{dt} = -k_1(S)_0(E) + k_{-1}(ES') + k_3(ES'') \tag{10.123}$$

$$\frac{d(ES')}{dt} = k_1(S)_0(E) - [k_{-1} + k_2](ES') \tag{10.124}$$

$$\frac{d(ES'')}{dt} = k_2(ES') - k_3(ES'') \tag{10.125}$$

$$\frac{d(P')}{dt} = k_2(ES') \tag{10.126}$$

These approximate differential equations are equivalent to those obtained by Ouellet and Stewart,[7] who began their derivation with a mechanism in which the reverse arrows were omitted from Reactions (10.103) and (10.104). This incomplete mechanism violates the principle of microscopic reversibility. The chemical model of Gutfreund and Sturtevant is self-consistent. In particular, it is consistent with the principle of microscopic reversbility. The approximations made by Gutfreund and Sturtevant are numerical approximations and as such can be justified or rejected in a specific situation in strictly numerical terms. The chemical model of Ouellet and Stewart contains approximations, inconsistent with microscopic reversbility, within the chemical model itself. It is not clear how to specify quantitatively the range of validity of their resulting mathematical equations, because the original approximations are neither algebraic nor numerical in nature.

We now assume solutions of the form

$$(E) = ae^{-mt} \qquad (ES') = be^{-mt} \qquad (ES'') = ce^{-mt} \qquad (P') = pe^{-mt}$$

and substitute these trial solutions into Equations (10.123)–(10.126). The result is, after division by the common exponential factor, e^{-mt}, the set of simultaneous equations

$$a[k_1(S)_0 - m] - b[k_{-1}] \qquad\qquad - c[k_3] \qquad\qquad = 0$$

$$-a[k_1(S)_0] \qquad + b[k_{-1} + k_2 - m] \qquad\qquad = 0$$

$$- b[k_2] \qquad + c[k_3 - m] \qquad = 0$$

$$- b[k_2] \qquad\qquad - p[m] = 0$$

[7] L. Ouellet and J. A. Stewart, *Can. J. Chem.*, **37**, 737 (1959).

The associated determinantal equation is

$$
\begin{vmatrix}
[k_1(S)_0 - m] & -[k_{-1}] & -[k_3] & 0 \\
-[k_1(S)_0] & [k_{-1} + k_2 - m] & 0 & 0 \\
0 & -[k_2] & [k_3 - m] & 0 \\
0 & -[k_2] & 0 & -[m]
\end{vmatrix} = 0
$$

which, when expanded, is the quartic equation

$$m^4 - m^3\{k_1(S)_0 + k_{-1} + k_2 + k_3\}$$
$$+ m^2\{k_1(S)_0[k_2 + k_3] + k_{-1}k_3 + k_2k_3\} = 0 \quad (10.127)$$

Two of the four roots are zero, and the other two are

$$m_1 + m_2 = k_1(S)_0 + k_{-1} + k_2 + k_3 \quad (10.128)$$
$$m_1 m_2 = k_1(S)_0[k_2 + k_3] + k_{-1}k_3 + k_2k_3 \quad (10.129)$$

If $m_1 \gg m_2$, then

$$m_1 = k_1(S)_0 + k_{-1} + k_2 + k_3 \quad (10.130)$$

$$\frac{m_1 m_2}{m_1 + m_2} = m_2 = \frac{k_1(S)_0[k_2 + k_3] + k_{-1}k_3 + k_2k_3}{k_1(S)_0 + k_{-1} + k_2 + k_3} \quad (10.131)$$

The concentration of product P' is given as a function of time, during the transient phase and subject to the approximations made in the derivation, by

$$(P') = p_0 + p_1 e^{-m_1 t} + p_2 e^{-m_2 t} \quad (10.132)$$

Because of the approximations made [(10.120), (10.121), and (10.122)], equations such as (10.132) will not converge properly at large values of t to the appropriate equilibrium concentrations. Therefore, Equation (10.132) cannot apply throughout the entire steady-state phase of the reaction.

When an experimental rate curve such as the one presented in Figure 9.1 is obtained, it can be analyzed by considering Equations (10.119) and (10.132) to be particular solutions of the complete set of differential equations, with a general initial-rate solution being the sum of the two equations. It is the judgment of the chemists who have studied α-chymotrypsin reactions that, under the conditions and with the instrumentation usually employed for investigation of these rates, m_1 is too large and the first relaxation thus too fast to have been observed. Then, considering the situation after the first relaxation has proceeded long enough so that the exponential term $p_1 e^{-m_1 t}$ has become insignificantly small, we can add together Equations (10.119) and (10.132) to get an equation of the form

$$(P') = Bt + Ce^{-m_2 t} + D \tag{10.133}$$

where

$$B = \frac{(E)_0 (S)_0 k_1 k_2 k_3}{(S)_0 k_1 [k_2 + k_3] + k_3 [k_{-1} + k_2]} \tag{10.134}$$

$$m_2 = \frac{(S)_0 k_1 [k_2 + k_3] + k_3 [k_{-1} + k_2]}{k_1 (S)_0 + k_{-1} + k_2 + k_3} \tag{10.135}$$

In actual experiments, there is a contribution to the rate of production of P′ from the spontaneous hydrolysis of the substrate. This small but measurable nonenzymatic reaction produces an absorbance change that can be measured independently in a control experiment. The absorbance change from the nonenzymatic reaction can be subtracted at each value of t from the absorbance observed in the presence of enzyme to yield the corrected absorbance curve. It is not necessary to make the correction if only m_2 is being determined, but it is necessary when B is being evaluated.

Determination of B and m_2 from Absorbance Measurements. It is possible to find a wavelength at which the only absorbance changes can be attributed to P′, and thus the absorbance of the solution can be written as

$$A = l\epsilon Bt + l\epsilon Ce^{-m_2 t} + E \tag{10.136}$$

where A is the instantaneous value of the absorbance; l is the optical path length through the reacting solution; ϵ is the absorptivity of the species P′; B, C, and m_2 have the same meanings as the symbols did in Equation (10.133); and the constant E may contain time-independent absorbance contributions from other species in the solution as well as the term $l\epsilon D$.

The value of B is determined by measuring the slope of the linear portion of the absorbance–time curve after the $Ce^{-m_2 t}$ terms has become essentially equal to zero. That slope is equal to $l\epsilon B$, and it is necessary to have independent knowledge of the values of l and ϵ.

The macroscopic rate constant m_2 can be evaluated conveniently by means of a modified Guggenheim method, described by Gutfreund and Sturtevant. We define the quantities

$$A_t = A \text{ at time equal to } t$$

$$A_{t+\tau} = A \text{ at time equal to } [t + \tau]$$

$$A_{t+2\tau} = A \text{ at time equal to } [t + 2\tau]$$

where the quantity τ is an arbitrary time interval chosen at the convenience of the experimenter, preferably several times greater than $[1/m_2]$. Equation (10.136) can be written at each of the times, yielding

$$A_t = l\epsilon Bt + l\epsilon Ce^{-m_2t} + E \tag{10.137}$$

$$A_{t+\tau} = l\epsilon B[t + \tau] + l\epsilon Ce^{-m_2[t+\tau]} + E \tag{10.138}$$

$$A_{t+2\tau} = l\epsilon B[t + 2\tau] + l\epsilon Ce^{-m_2[t+2\tau]} + E \tag{10.139}$$

Now calculate the quantity $(A_t - 2A_{t+\tau} + A_{t+2\tau})$ by adding and subtracting Equations (10.137)–(10.139):

$$(A_t - 2A_{t+\tau} + A_{t+2\tau})$$

$$= l\epsilon Bt + l\epsilon Ce^{-m_2t} + E$$

$$- 2l\epsilon B[t + \tau] - 2l\epsilon Ce^{-m_2[t+\tau]} - 2E$$

$$+ l\epsilon B[t + 2\tau] + l\epsilon Ce^{-m_2[t+2\tau]} + E$$

$$= l\epsilon C\{e^{-m_2t} - 2e^{-m_2[t+\tau]} + e^{-m_2[t+2\tau]}\}$$

$$= l\epsilon C\{e^{-m_2t} - 2e^{-m_2t}e^{-m_2\tau} + e^{-m_2t}e^{-2m_2\tau}\}$$

$$= l\epsilon Ce^{-m_2t}\{1 - 2e^{-m_2\tau} + e^{-2m_2\tau}\}$$

$$= \{\text{constant}\}e^{-m_2t} \tag{10.140}$$

The logarithm to the base e is taken of both sides of Equation (10.140), giving

$$\ln | A_t - 2A_{t+\tau} + A_{t+2\tau} | = \text{constant} - m_2t \tag{10.141}$$

Equation (10.141) is the equation of a straight line with slope equal to $-m_2$.

It has been shown, for a suggested mechanism for α-chymotrypsin reactions, how transient-phase rate measurements can supplement steady-state data for the same reaction.

The complete set of differential equations for the mechanism was written. It was not possible to obtain an exact solution valid throughout the entire reaction.

There are certain numerical approximations which, if valid, permit solution of the differential rate equations. Such approximations were made, and two approximate solutions were thereby obtained. We found a transient-phase solution and a steady-state solution. We must remember that the validity of each approximate solution depends on the validity of the original numerical approximations.

It may be possible to find reaction conditions under which, to within an approximation that involves errors no greater than the uncertainties in the experimental rate data, the transient-phase solution and the steady-state initial rate solution can be considered to be particular solutions that are valid in overlapping time intervals. A general solution—in this case (10.133)—can then be obtained by adding the two particular solutions.

Transient-phase rate measurements can yield quantitative information about reaction mechanisms which is otherwise inaccessible to chemists. It is essential that experiments be designed and interpreted with due regard to the limitations and restrictions imposed by those numerical approximations which were required to solve the differential equations.

Problems

10.1. It is asserted in this chapter that inequalities (10.10) require that m_1 be much greater than m_2. Assume some appropriate numerical values of the four microscopic rate constants, and show that for this specific case the two m's have very different numerical values. Can you demonstrate algebraically that this is a general result?

10.2. Verify, for Mechanism (10.26), that $d(S)/dt = -d(P)/dt$. Under what conditions are these two derivatives equal for Mechanism (10.38)–(10.39)? Can these two derivatives be equal for a similar mechanism with n intermediates?

10.3. Show how each of the microscopic rate constants for Mechanism (10.38)–(10.39) can be evaluated from the values of the steady-state macroscopic kinetic constants V_S, V_P, K_S, and K_P.

10.4. Demonstrate that, in Equation (10.133), $D = -C$.

10.5. Show that the principle of microscopic reversibility requires that

$$\frac{V_S K_P}{V_P K_S} = K'$$

for Mechanism (10.38)–(10.39). The quantity K' is the overall apparent equilibrium constant for the reaction. What is the analogous relationship for Mechanism (10.85)–(10.87)?

10.6. Many enzymes have an amazing substrate specificity, sometimes catalyzing the reaction of only a single substrate–product pair. However, certain related compounds may combine with the enzyme, thus reducing the number of catalytically active enzyme molecules available for reaction with the substrate. This is a case of *competitive inhibition* by the inhibitor, I. One mechanism that demonstrates competitive inhibition is

$$E + S \underset{k_{-1}}{\overset{k_1}{\rightleftharpoons}} X$$

$$X \underset{k_{-2}}{\overset{k_2}{\rightleftharpoons}} E + P$$

$$E + I \underset{k_{-3}}{\overset{k_3}{\rightleftharpoons}} EI$$

Show how this mechanism predicts an initial-rate, steady-state equation of the form

$$\frac{d(P)}{dt} = \frac{V_S}{1 + \dfrac{K_S}{(S)_0}\{1 + K_I(I)_0\}}$$

How can K_I be evaluated?

10.7. For Mechanism (10.38)–(10.39), find the maximum attainable steady-state initial rate, starting with substrate S. This will be the limiting form of Equation (10.83) at high values of $(S)_0$. What is the rate when $(S)_0 = K_S$?

Bibliographical Note

It is often difficult to justify the steady-state assumption for a particular reaction. Some criteria are developed by R. M. Noyes ("Kinetic Treatment of Consecutive Processes," in G. Porter, ed., *Progress in Reaction Kinetics*, vol. 2, New York: The Macmillan Company, 1964) for the validity of what he calls the rate-determining step approximation, the uniform-flux approximation, and the first-order sequence approximation.

An excellent pair of books for more intensive study of enzymatic catalysis is C. Walter, *Steady-State Applications in Enzyme Kinetics*, New York: The Ronald Press Company, 1965; and C. Walter, *Enzyme Kinetics: Open and Closed Systems*, New York: The Ronald Press Company, 1966.

The uses of relaxation methods in studying enzymatic reactions, with particular emphasis on the possibilities of elucidating the chemistry of elementary reaction steps, are reviewed in G. G. Hammes, *Accounts of Chem. Res.*, **1,** 321 (1968).

The history of catalysis is filled with contradictory hypotheses and theories, with conflict and paradox. The development of some of the central ideas of chemical catalysis is traced by V. I. Kuznetsov (*Chymia*, vol. 11, Philadelphia: University of Pennsylvania Press, 1966, p. 179, trans. from Russian by H. M. Leicester).

11

Irreversibility

Microscopic reversibility is a fundamental feature of elementary molecular chemical reactions. Viewed at the molecular level, all chemical interaction and all chemical change are reversible. An animated motion picture film can be produced to show the changes in individual molecules during an elementary reaction, and there is no way for a viewer to tell whether the film is being run forward or backward through the motion picture projector. The variable t enters into the description of chemical reaction dynamics of individual molecules in such a way that changing the sign of t has no substantive effect on the processes.

Reversibility is woven into the fabric of our descriptions of elementary chemical processes. Nevertheless, there is an equally fundamental *irreversibility* inherent in our macroscopic descriptions of the same processes when very large numbers of molecules are simultaneously involved. An isolated nonequilibrium system evolves toward equilibrium. An equilibrium system in isolation does not drift away from equilibrium. The sign of t is important in describing the evolution of a system toward equilibrium. Time is a vector, an arrow, which points forward in the direction—for an isolated system containing a very large number of molecules—of equilibrium.

In isolation, or in a suitably controlled constant environment, the macroscopic future of a chemically reacting system is completely determined by its present nonequilibrium state. For every reaction mechanism we have examined there is a set of simultaneous differential equations that describes the time course of that chemical reaction. To make specific predictions from these differential equations for a specific reaction system, it is only necessary to specify the numerical values of the concentrations at some particular value of t. Then the concentrations at all future times are in principle calculable. The evolution in time of each of the concentration variables that describe the instantaneous macroscopic state of the chemical system is quantitatively predetermined for all the future.

Information about the macroscopic history of a reacting system is continuously destroyed as the system approaches the state of chemical equilibrium. A variety of different initial states *might* have reacted to give the present state. There is an unlimited number of possible initial composition states which would have yielded the same present state. Yet no information is contained within the present state of the chemical system to tell which of the many conceivable initial states was the actual initial state. The reacting system contains no memory of its own unique past. The system moves irreversibly through time toward a predestined future without leaving a record of the particular details of its past.

Physical chemists and chemical physicists who are concerned with the theoretical foundations of chemical reaction kinetics have been asking some searching questions about the approach to equilibrium. There continue to be reports of chemical reactions which, instead of approaching an equilibrium point in a monotonic manner, exhibit periodic behavior.[1] Observed periodicity raises some fundamental questions. Does a chemical reaction necessarily, regardless of mechanism, follow an irreversible path that leads to equilibrium? Is the reaction path uniquely determined by initial conditions? Is there one unique equilibrium point for each particular chemical system? Can there be oscillations of the entire chemical system about an equilibrium point? How are answers to these questions modified for the

---

[1] An interesting historical survey of periodic phenomena in chemical kinetics is found in E. S. Hedges and J. E. Myers, *The Problem of Physico-Chemical Periodicity*, London: Edward Arnold & Co., 1926; Xerox reprint available from University Microfilms, Ann Arbor, Mich. About a third of the book is devoted to time periodicity in chemical reactions, mostly reactions taking place in heterogeneous systems. There is evidence that the reduction of iodic acid with hydrogen peroxide is oscillatory [W. C. Bray, *J. Amer. Chem. Soc.*, **43**, 1262 (1921); F. O. Rice and O. M. Reiff, *J. Phys. Chem.*, **31**, 1352 (1927)], with a brown color due to free iodine appearing and disappearing with a frequency that depends on the pH and the temperature, and with evolution of oxygen gas in gusts. Sustained oscillations of the concentrations of fructose-6-phosphate, fructose-6-diphosphate, and other intermediates in glucose metabolism, in the presence of virtually constant glucose concentration, have been observed in suspensions of yeast cells [J. Higgins, *Proc. Nat. Acad. Sci. U. S.*, **51**, 989 (1964); B. Chance, R. W. Estabrook, and A. Ghosh, *Proc. Nat. Acad. Sci. U.S.*, **51**, 1244 (1964); A. Ghosh and B. Chance, *Biochem. Biophys. Res. Commun.*, **16**, 174 (1964)]. A solution composed of malonic acid, ceric sulfate, potassium bromate, and sulfuric acid in water has an absorbance at 317 mμ which is periodic in time [H. Degn, *Nature*, **213**, 589 (1967); see also *Chemistry*, **41** (5), 26 (1968)]. Some progress has been made in formulating a theory of oscillating chemical reactions; see J. Higgins, *Ind. Eng. Chem.*, **59** (5), 18 (1967), reprinted as chapter 12 of *Applied Kinetics and Chemical Reaction Engineering*, Washington, D.C.: American Chemical Society, 1967.

special cases of a system of just a few molecules, an isolated system, a closed system in a thermostat, or an open system with some reactants and products continuously exchanging with surroundings (as for example, in an industrial flow reactor, or in a living cell)?

The answers probably do not lie within the sets of linear differential equations that we have been investigating. All the systems of linear differential equations that have been considered in this book are well behaved in terms of having an asymptotic approach to a unique infinite-time value for each of the concentrations. The existence of a periodic solution certainly requires nonlinear differential rate equations.[2] The conditions that must be met by a kinetic reaction mechanism to guarantee that there will be a unique equilibrium point and that the reaction will converge to that equilibrium point have been examined by Wei and Prater.[3] For chemical reaction systems, a well-behaved convergence to equilibrium is assured if it can be shown that, when a reaction path of the mechanism passes through a composition point P (point P representing the concentration of each of the reactive species in the system), there is then a neighborhood about point P to which the reaction path will never again return.

The linear differential rate equations for coupled unimolecular or pseudo-unimolecular reactions require irreversible evolution of the reacting system toward a unique equilibrium point. For these mechanisms, irreversibility enters the description of reaction rates somewhere between the quantum-mechanical description of the molecules as individual entities and the formulation of the macroscopic mathematical model in terms of molar concentrations. The irreversible evolution toward equilibrium is related to the existence of very large numbers of molecules in the reacting system. This is a research problem being worked on in the field of chemical statistical mechanics. The question of how a *moderate* number of molecules behave and how the evolution of a moderate number of reacting molecules can be described remains at present unresolved.

The final approach to equilibrium by a reacting system that is already near equilibrium can be described in terms of thermodynamic forces and an accompanying entropy production. Such a discussion takes place within the context of the theories of irreversible thermodynamics.[4] This approach

---

[2] T. A. Bak, *Contributions to the Theory of Chemical Kinetics*, New York: W. A. Benjamin, Inc., 1963, chap. 3; R. Lefever, G. Nicolis, and I. Prigogine, *J. Chem. Phys.*, **47**, 1045 (1967).

[3] J. Wei and C. D. Prater, in *Advances in Catalysis*, vol. 13, New York: Academic Press, Inc., 1962, pp. 343ff.; J. Wei, *J. Chem. Phys.*, **36**, 1578 (1962).

[4] The triangular mechanism (5.84)–(5.86) is discussed in such a context by T. L. Hill (*Thermodynamics for Chemists and Biologists*, Reading, Mass.: Addison-Wesley Publishing Company, Inc., 1968) in his chapter, "Introduction to Irreversible Thermodynamics," p. 156. A parallel discussion of the same mechanism to illustrate the Onsager reciprocal relations is given by K. G. Denbigh (*The Thermodynamics of the Steady State*, London: Methuen & Co., Ltd, 1951) in his chapter, "Onsager's Theory," p. 31.

to equilibrium can also be meaningfully related to increasing disorder in the chemical system.[5]

SUGGESTED DEMONSTRATION

A simple demonstration makes plausible the statement that irreversibility can be a consequence of the fact that the system contains a very large number of molecules.

Place two marbles, of different colors, in a tray. Shake this tray gently, and observe what happens.

Place 50 marbles of the same color on one side of a tray, and place 50 marbles of another color on the opposite side of the same tray. Shake this tray gently, and observe what happens.

The tray with 100 marbles begins with a "nonequilibrium" distribution of marbles. Randomizing occurs as a result of the shaking, during which the collection tends toward the expected confusion of an "equilibrium distribution."

Each of the many possible distributions of the marbles is equally probable, so why is it that we would say that the initial distribution of the two colors separated is "nonequilibrium," while the confused mixture of colors that almost surely results from the shaking operation is "equilibrium"?

BIBLIOGRAPHICAL NOTE

The statement that "a chemical system cannot evolve *away* from equilibrium is valid only when the number of chemical entities in the system is very large. An effective and convincing presentation of this fact is the computer-produced film, B. J. Alder and F. Reif, *Irreversibility and Fluctuations* (8-mm cartridged film loops, black and white), New York: McGraw-Hill, Inc., 1967. The film shows calculated motions of particles inside a two-dimensional box, illustrating the approach to equilibrium, irreversible behavior, and fluctuations in the equilibrium state for systems with just a few particles and for systems with many particles.

[5] Nobel Laureate George Porter presented a series of British Broadcasting Corporation lectures on "The Laws of Disorder." *Chemistry* published the lectures, beginning in May 1968. See especially "Entropy and the Second Law," *Chemistry*, **41** (7), 36 (1968); "Equilibrium—The Limit of Disorder," *ibid.*, **41** (8), 16 (1968); "Times of Change," *ibid.*, **41** (10), 24 (1968). T. R. Blackburn [*Equilibrium: A Chemistry of Solutions*, New York: Holt, Rinehart and Winston, Inc., 1969] introduces students to chemical equilibria in solutions by considering the principle of ever-increasing microscopic disorder in his chapter 1, "Order out of Chaos."

12

Encounter, Activation, Transition, and Reaction

A chemist can ask a virtually unlimited number of questions about the details of molecular interactions and molecular rearrangements involved in an elementary reaction such as

$$A + B \underset{k_{-1}}{\overset{k_1}{\rightleftharpoons}} C + D \tag{12.1}$$

Much of the thrust of chemical theory is directed toward understanding the events associated with making and breaking chemical bonds, with the intimate details of the transformation of reactant molecules into product molecules. Theories of chemical dynamics that attempt to elucidate the molecular course of a reaction in terms of the fundamentals of quantum theory are incomplete. This is a frontier area of chemical research, and it is likely that for many decades the number of questions asked will easily surpass the number of answers obtained.

Certain features of theory are well established. It is clear that ordinarily Reaction (12.1), even though called an elementary reaction, must occur in a sequence of stages. If A and B are to react chemically in any way, they must come reasonably close together. There must be an *encounter*. For each chemical reaction there is a minimum chemical interaction distance; molecules farther apart cannot react, although there may be mutual electrostatic attraction or repulsion at these greater distances. The requirement that initially separated molecules must come within a minimum interaction distance in order to form an encounter pair can be formulated as

$$A + B \rightleftharpoons A \cdots B \rightleftharpoons C \cdots D \rightleftharpoons C + D \qquad (12.2)$$

It is difficult to visualize the process of chemical transformation without introducing still another species into this formulation. A molecule having a geometric and electronic structure intermediate between the structures of the two encounter pairs almost surely is needed as an intermediate in this chemical transition. We shall call this species X a *transition complex*. Thus the minimum sequence of molecular events which seems needed to account for elementary reaction (12.1) becomes

$$A + B \rightleftharpoons A \cdots B \rightleftharpoons X \rightleftharpoons C \cdots D \rightleftharpoons C + D \qquad (12.3)$$

By the very nature of being intermediate in structure between two stable chemical entities, the transition complex is likely to be unstable. Formation of X may require enough extra energy, the *activation energy*, to qualify the transition complex for the name of an *activated complex*. Theories have been developed that make possible an interpretation of the experimentally determined temperature dependence of a microscopic rate constant in terms of the activation energy for the associated elementary reaction.

Not all chemical reactions fit neatly into the conceptual scheme we are developing. But certainly this scheme is very useful in organizing information about the molecular dynamics of chemical reactions. This scheme has been fruitful in suggesting experiments that point to the fundamentals of chemical dynamics. And this scheme has been a means of focusing the work of those chemists who devote much of their time to theoretical calculations and investigations in the broad field of chemical dynamics.

Encounter. A dilute solution is almost entirely composed of solvent molecules. When a molecule of A, in its random movement among solvent molecules, comes within the requisite chemical interaction distance of molecule B, this encounter pair will be surrounded by many solvent molecules, enclosed in a cage of solvent. Ordinarily, having once come together, this encounter pair will stay together for awhile.

When a molecule of A and a molecule of B are so close that they are deforming one another, they are so close that there is a net repulsive force between the two molecules. The molecules are in a state of collision. A collision may, or may not, result in formation of the transition complex X. Two molecules are in a state of collision for about 3×10^{-13} second, and the time between collisions is of comparable magnitude. An encounter in solution has a lifetime that is often in the range 10^{-10} to 10^{-8} second, longer in the case of viscous liquids, and of the order of several seconds to days for glasses.[1] Typically, one can expect many collisions between the two reactant molecules while they are trapped within the solvent cage during a single encounter. In addition, each reactant molecule undergoes many collisions with solvent molecules during this encounter period, and each such collision involves some interchange of energy between reactant and solvent molecules. A fortuitous series of collisions between reactant and solvent can yield a reactant molecule that has, for a short time, an energy much greater than the average energy of similar molecules throughout the solution.

An encounter can be concluded in two different ways. One of the potential reactants can escape from the cage, or there can be a *reactive collision*. These two possibilities are symbolized by two of the arrows in the scheme in (12.3).

Conditions for a Reactive Collision.   For some reactions, it is highly probable that one of the first collisions during an encounter will result in reaction to give the transition complex X. With other reactions, the probability of a collision leading to X is very small. This difference in the probability that a given collision will be a reactive collision has been interpreted as being due in part to a difference between the energy of the average encounter pair, A$\cdots$B, and the energy of the average transition complex, X. This probability difference is also due in part to geometric requirements imposed upon any collision complex that is to become X.

As a rough approximation, any geometric or orientation requirement can be considered to reduce the probability of reaction by a constant, temperature-independent factor. The difference in the average energy between the encounter pair and the transition complex reduces the probability of reaction by the factor

$$\exp\left(-\left[\Delta E_{\text{activation}}/RT\right]\right) \tag{12.4}$$

where $\Delta E_{\text{activation}}$ is the activation energy per mole. The change in numerical value of this exponential factor arising from a change of a few degrees in the temperature can be insignificant or massive, depending on the value of $\Delta E_{\text{activation}}$. Table 12.1 shows the effect on the probability of reaction of a 10°C temperature change. There is no temperature effect on the reaction probability if $\Delta E_{\text{activation}}$ is zero, and very little effect if $\Delta E_{\text{activation}}$ is less than 1 kcal per mole. When the activation energy is about 10 kcal per mole, a 10-degree temperature rise increases the reaction probability by almost

---

[1] A. M. North, *The Collision Theory of Chemical Reactions in Liquids*, New York: John Wiley & Sons, Inc., 1964, pp. 7, 59.

TABLE 12.1.   Exponential Reaction-Probability Factor

$\Delta E_{activation}$ (cal/mole)	$\exp(-[\Delta E_{activation}/RT])$		Ratio of the Exponential Evaluated at $T = 300°K$ and $T = 310°K$
	$T = 300°K$	$T = 310°K$	
0	1	1	1
100	0.846	0.850	1.005
1 000	0.187	0.197	1.055
10 000	4.11×10^{-6}	7.06×10^{-6}	1.716
100 000	1.48×10^{-73}	3.26×10^{-71}	221

a factor of two. At higher activation energies, the probability of reaction becomes fantastically small, and the temperature dependence becomes very large.

Reaction.   The elementary process

$$A + B \rightarrow C + D$$

has not been completed until X changes into the complex $C \cdots D$ and then dissociates into separately solvated C and D species. The sequence of events during this elementary process can be visualized as progress along a multi-barrier pathway, a pathway that has come to be known as the reaction coordinate. Figure 12.1 shows a useful representation of the energy relations in this elementary process. There are four energy barriers between separated reactants and separated products. The reacting chemical species must obtain energy from the surrounding solvent molecules in order to surmount each barrier. The principle of microscopic reversibility requires not only that there be a reverse process

$$A + B \leftarrow C + D$$

but also that there be detailed balancing with respect to passage over each of the energy barriers. At equilibrium, the number of species passing over an energy barrier in one direction must be equal to the number of species passing over from the reverse direction, during any particular period of time.

Figure 12.1 is a profile of what must be a multidimensional potential-energy surface. Potential-energy surfaces in three dimensions (in three-dimensional configurational space) have been calculated for a three-particle

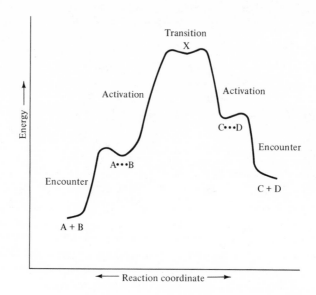

FIGURE 12.1. A reaction profile

system,[2] and then mapped and interpreted as a reaction landscape. Construction of even approximate potential-energy surfaces to describe most chemical reactions in solution, perhaps possible in principle, are beyond the present power of chemical quantum theory. The best that can be done is the sketching of the reaction profile, and even in these two dimensions the challenges involved in coming to reasonable and justifiable estimates of the heights of energy barriers can be impressive.

Experimental Evaluation of Activation Energy. For every energy barrier there is an activation energy and a reaction probability with the factor (12.4). It can be plausibly argued that an overall reaction probability, P, for an elementary process should be the product of the individual barrier reaction probabilities, so that

$$P = Ce^{-[\Delta E_1/RT]}e^{-[\Delta E_2/RT]}e^{-[\Delta E_3/RT]}\ldots$$

$$= Ce^{-[\Delta E_1+\Delta E_2+\Delta E_3+\cdots]/RT}$$

$$= Ce^{-\Delta E/RT} \tag{12.5}$$

    [2] H. Eyring and E. M. Eyring, *Modern Chemical Kinetics*, New York: Van Nostrand Reinhold Company, 1963, pp. 10–28, 77–80. See also a paper in M. H. Back and K. J. Laidler, *Selected Readings in Chemical Kinetics*, New York: Pergamon Press, Inc., 1967: H. Eyring and M. Polanyi, "On Simple Gas Reactions," trans. from *Z. Physik. Chem.* (Leipzig), **B12**, 279 (1931).

where C is a temperature-independent constant and $\Delta E = \Delta E_1 + \Delta E_2 + \Delta E_3 + \cdots$ is the overall activation energy for the elementary process. It is this sort of reasoning that can lead to Equation (3.6),

$$k = Ze^{-\Delta E/RT}$$

and can persuade us to identify the quantity ΔE in Equation (12.5) with the quantity ΔE in Equation (3.6). In logarithmic form, Equation (3.6) is

$$\ln k = \ln Z - \frac{\Delta E}{RT} \tag{12.6}$$

The experimental activation energy can be directly evaluated by determining the value of a microscopic rate constant at each of several values of the temperature. Then a plot of $\ln k$ versus $[1/T]$ can be constructed. This plot should be a straight line, with slope equal to $-\Delta E/R$.

Absolute Rate Constants and Relative Rate Constants. The a priori calculation of the numerical value of any rate constant for a reaction in solution is difficult. There has been success for certain limiting cases. For example, when virtually every encounter leads to reaction, the reaction rate is limited by the rate at which reactants can diffuse to within the encounter distance. Such a reaction is said to be *diffusion controlled*, and its rate can be estimated by means of diffusion theory.

Interest continues to be great in semiempirical theories that deal with predictions and correlations of values of rate constants for rather complicated reactions in solution. Here the successes have come largely with efforts to *compare* rate constant values, to correlate *relative* values of rate constants. Detailed theories have been developed to deal with the effect of altering the solvent composition, or changing the ionic environment, on the rates of reactions.[3]

Vigorous and imaginative research continues to provide more and more information about reaction mechanisms and about the molecular structure of reactants, intermediates, and products involved in reaction sequences. As more mechanisms are elucidated, the opportunities become greater for making useful and meaningful comparisons and correlations,

[3] Substantial sections of the following four books are devoted to just such correlations:

E. S. Amis, *Kinetics of Chemical Change in Solution*, New York: The Macmillan Company, 1949.

E. S. Amis, *Solvent Effects on Reaction Rates*, New York: Academic Press, Inc., 1966.

R. P. Bell, *Acid–Base Catalysis*, New York: Oxford University Press, 1941.

E. M. Kosower, *An Introduction to Physical Organic Chemistry*, New York: John Wiley & Sons, Inc., 1968.

and for attempting some wide-ranging generalizations. Textbooks, notably in physical organic chemistry but also in biochemistry and inorganic chemistry, are being written with unifying themes based on semiempirical correlations of rate constants and mechanisms for significant series of chemical reactions. Such correlations are bringing important insights into chemistry.

Comparisons of mechanisms depend on the validity and credibility of the mechanisms themselves. Because of the very nature of reaction mechanisms, any particular mechanism is necessarily tentative. Any mechanism is subject to revision, modification, or replacement. A mechanism proposed for a reaction that has been experimentally investigated only casually, or with inadequate chemical, instrumental, or mathematical tools, is especially subject to being changed when new experiments are performed. Some energetic and creative chemists will seek out and study critical chemical reactions with inadequately substantiated mechanisms. Such studies have had in the past, and will surely have in the future, a dynamic impact on chemical theories.

Appendix I
Use of Determinants to Solve Simultaneous Equations

Consider the set of simultaneous equations

$$a_{11}x_1 + a_{12}x_2 = 0 \qquad (I.1)$$

$$a_{21}x_1 + a_{22}x_2 = 0 \qquad (I.2)$$

The coefficients a_{11} (read a-one-one), a_{12}, a_{21}, and a_{22} are constants, and x_1 and x_2 are variables. We wish to find the relationship that must hold among the a's so that the two equations will be simultaneously valid for arbitrary values of the x's. The approach to be used is based on the matrix equation

$$\begin{pmatrix} a_{11} & a_{12} \\ a_{21} & a_{22} \end{pmatrix} \begin{pmatrix} x_1 & 1 \\ x_2 & 1 \end{pmatrix} = \begin{pmatrix} [a_{11}x_1 + a_{12}x_2] & [a_{11} + a_{12}] \\ [a_{21}x_1 + a_{22}x_2] & [a_{21} + a_{22}] \end{pmatrix} \qquad (I.3)$$

Matrix equation (I.3) is very closely related to the set of simultaneous algebraic equations (I.1)–(I.2). Being written in terms of matrices, however, the matrix equation may not look at all familiar. So before proceeding we shall discuss a few relevant features and properties of matrices.

Some Properties of Matrices. A matrix is an orderly arrangement of numbers displayed in a rectangular table. The individual entries, called *elements*, are lined up in rows and columns. To identify an element with

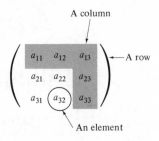

A column

A row

An element

FIGURE I.1. A 3 × 3 matrix.

its location in the matrix, it is convenient to use a letter with two sub-scripts to symbolize the element. The first subscript indicates the number of the row, and the second indicates the number of the column.

To interpret the indicated product of two matrices in Equation (I.3), matrix multiplication must be defined. Multiplication of a 2 × 2 matrix (read two-by-two matrix, meaning a matrix with two rows and two columns) by a second 2 × 2 matrix yields a third 2 × 2 matrix. Multiplication of two 3 × 3 matrices results in

$$
\begin{pmatrix} a_{11} & a_{12} & a_{13} \\ a_{21} & a_{22} & a_{23} \\ a_{31} & a_{32} & a_{33} \end{pmatrix}
\begin{pmatrix} b_{11} & b_{12} & b_{13} \\ b_{21} & b_{22} & b_{23} \\ b_{31} & b_{32} & b_{33} \end{pmatrix}
=
\begin{pmatrix} c_{11} & c_{12} & c_{13} \\ c_{21} & c_{22} & c_{23} \\ c_{31} & c_{32} & c_{33} \end{pmatrix}
$$

The new element c_{11} is obtained from row 1 of the a matrix and column 1 of the b matrix, using the equation

$$
c_{11} = a_{11}b_{11} + a_{12}b_{21} + a_{13}b_{31}
$$

In general, element c_{ij} is obtained from row i of the a matrix and column j of the b matrix by summing three row–column products:

$$
c_{ij} = a_{i1}b_{1j} + a_{i2}b_{2j} + a_{i3}b_{3j}
$$

The procedure is schematically represented in Figure I.2. Matrix multiplication is an orderly and straightforward procedure even for larger matrices. For multiplication of an $n \times n$ matrix, the element c_{ij} is found from row i of the a matrix and column j of the b matrix according to

$$
c_{ij} = a_{i1}b_{1j} + a_{i2}b_{2j} + a_{i3}b_{3j} + \cdots + a_{in}b_{nj}
$$

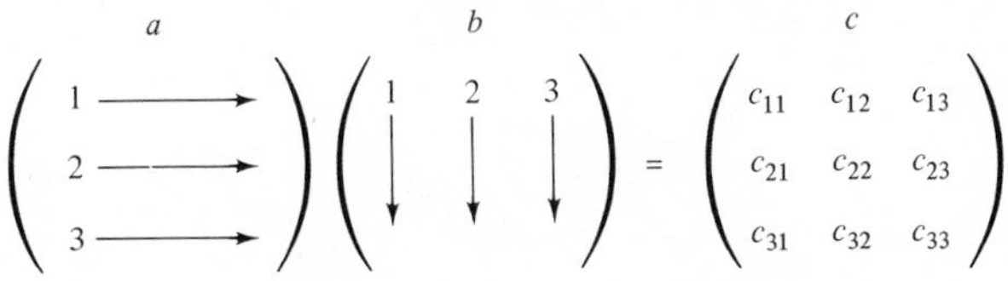

$$\begin{pmatrix} 1 \longrightarrow \\ 2 \longrightarrow \\ 3 \longrightarrow \end{pmatrix} \begin{pmatrix} 1 & 2 & 3 \\ \downarrow & \downarrow & \downarrow \\ & & \end{pmatrix} = \begin{pmatrix} c_{11} & c_{12} & c_{13} \\ c_{21} & c_{22} & c_{23} \\ c_{31} & c_{32} & c_{33} \end{pmatrix}$$

$$\overset{a}{} \qquad \overset{b}{} \qquad \overset{c}{}$$

FIGURE I.2. Schematic representation of matrix multiplication.

Associated with every square matrix is a number called the determinant of the matrix. Associated with the 2×2 matrix

$$\begin{pmatrix} a_{11} & a_{12} \\ a_{21} & a_{22} \end{pmatrix}$$

is the 2×2 determinant

$$\begin{vmatrix} a_{11} & a_{12} \\ a_{21} & a_{22} \end{vmatrix} \equiv a_{11}a_{22} - a_{21}a_{12} \qquad (I.4)$$

Associated with the 3×3 matrix

$$\begin{pmatrix} a_{11} & a_{12} & a_{13} \\ a_{21} & a_{22} & a_{23} \\ a_{31} & a_{32} & a_{33} \end{pmatrix}$$

is the 3×3 determinant

$$\begin{vmatrix} a_{11} & a_{12} & a_{13} \\ a_{21} & a_{22} & a_{23} \\ a_{31} & a_{32} & a_{33} \end{vmatrix} = a_{11} \begin{vmatrix} a_{22} & a_{23} \\ a_{32} & a_{33} \end{vmatrix} - a_{12} \begin{vmatrix} a_{21} & a_{23} \\ a_{31} & a_{33} \end{vmatrix} + a_{13} \begin{vmatrix} a_{21} & a_{22} \\ a_{31} & a_{32} \end{vmatrix}$$

$$= a_{11}[a_{22}a_{33} - a_{23}a_{32}] - a_{12}[a_{21}a_{33} - a_{23}a_{31}]$$

$$+ a_{13}[a_{21}a_{32} - a_{22}a_{31}]$$

The 3×3 determinant was evaluated by expansion in terms of three 2×2 determinants, followed by individual evaluation of the 2×2 determinants.

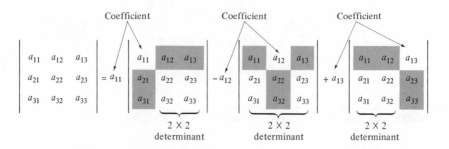

FIGURE I.3. Schematic representation of the expansion of a 3 × 3 determinant

The coefficients of the 2 × 2 determinants are the elements of the first row of the 3 × 3 determinant. The determinant associated with a particular coefficient is the array of elements remaining after the elements in the same row as the coefficients, and the elements in the same column as the coefficient, have been crossed off. The signs of the coefficients alternate, beginning with + for a_{11}, then − for a_{12}, and finally + for a_{13}. The final result of this stepwise expansion is an ordinary algebraic expression.

A 4 × 4 determinant can be expanded in the same fashion. The coefficients of the resulting 3 × 3 determinants will be the elements of the top row, with coefficients alternating in sign according to the sequence +, −, +, −. Then each of the 3 × 3 determinants of the first stage of this expansion is in turn expanded to 2 × 2 determinants. Finally, the collection of 2 × 2 determinants, each determinant evaluated by means of Equation (I.4), is expressed as an ordinary algebraic expression.

The definition of matrix multiplication and the definition of the determinant of a matrix together lead to a fundamental theorem which states that the determinant of the product of two matrices is equal to the product of the determinants of those matrices.

Return to the Simultaneous Equations. Equation (I.3) is the result of applying the rules for matrix multiplication to the left side of the equation. Two of the elements of the resulting matrix are seen to be identical to the left-hand members of Equations (I.1) and (I.2), and are thus each equal to zero. Equation (I.3) is therefore

$$\begin{pmatrix} a_{11} & a_{12} \\ a_{21} & a_{22} \end{pmatrix} \begin{pmatrix} x_1 & 1 \\ x_2 & 1 \end{pmatrix} = \begin{pmatrix} 0 & [a_{11} + a_{12}] \\ 0 & [a_{21} + a_{22}] \end{pmatrix}$$

The determinant of the product of two matrices is the product of the determinants of those matrices; thus

$$\begin{vmatrix} a_{11} & a_{12} \\ a_{21} & a_{22} \end{vmatrix} \begin{vmatrix} x_1 & 1 \\ x_2 & 1 \end{vmatrix} = \begin{vmatrix} 0 & [a_{11} + a_{12}) \\ 0 & [a_{21} + a_{22}] \end{vmatrix}$$

Evaluation of two of the determinants leaves

$$\begin{vmatrix} a_{11} & a_{12} \\ a_{21} & a_{22} \end{vmatrix} [x_1 - x_2] = 0 \tag{I.5}$$

We seek a relationship among the a's such that (I.1) and (I.2) will be valid for arbitrary values of x_1 and x_2, so Equation (I.5) requires that

$$\begin{vmatrix} a_{11} & a_{12} \\ a_{21} & a_{22} \end{vmatrix} = 0 \tag{I.6}$$

Evaluation of the determinant gives the result

$$a_{11}a_{22} - a_{12}a_{21} = 0$$

as the necessary condition for a simultaneous solution for the pair of algebraic equations.

The procedure for solving a pair of simultaneous equations—for finding the necessary relationship among the coefficients of the variables to permit a simultaneous solution for arbitrary values of those variables—is to form the determinantal equation analogous to (I.6). The first column contains the coefficients of one variable, and the second column contains the coefficients of the other variable. Coefficients from the same equation occupy the same row.

The procedure we have just developed is applicable to systems of simultaneous equations containing many equations and many unknowns. The case of three linear equations in three unknowns will be set up as an illustration. These three equations are

$$a_{11}x_1 + a_{12}x_2 + a_{13}x_3 = 0$$

$$a_{21}x_1 + a_{22}x_2 + a_{23}x_3 = 0$$

$$a_{31}x_1 + a_{32}x_2 + a_{33}x_3 = 0$$

If these three equations are to be simultaneously valid for arbitrary values of x_1, x_2, and x_3, the following determinantal equation must be satisfied by the nine coefficients:

$$\begin{vmatrix} a_{11} & a_{12} & a_{13} \\ a_{21} & a_{22} & a_{23} \\ a_{31} & a_{32} & a_{33} \end{vmatrix} = 0 \qquad (I.7)$$

Expansion of Equation (I.7), first to three 2×2 determinants, and then to an ordinary algebraic equation, gives the required relationship among the various coefficients.

Problems

I.1. Perform the indicated matrix multiplications:

a. $\begin{pmatrix} 1 & 2 & 3 \\ 4 & 5 & 6 \\ 7 & 8 & 9 \end{pmatrix} \begin{pmatrix} 1 & 1 & 1 \\ 1 & 1 & 1 \\ 1 & 1 & 1 \end{pmatrix} = ?$

b. $\begin{pmatrix} 1 & 2 & 3 \\ 4 & 5 & 6 \\ 7 & 8 & 9 \end{pmatrix} \begin{pmatrix} 1 & 0 & 0 \\ 0 & 1 & 0 \\ 0 & 0 & 1 \end{pmatrix} = ?$

I.2. Evaluate the following determinants, giving the answer in each case as a numeral:

a. $\begin{vmatrix} 2 & 4 & 6 \\ 1 & 2 & 3 \\ 7 & 8 & 9 \end{vmatrix} = ?$ b. $\begin{vmatrix} 1 & 1 \\ 1 & 1 \end{vmatrix} = ?$ c. $\begin{vmatrix} 1 & 0 \\ 0 & 1 \end{vmatrix} = ?$

d. $\begin{vmatrix} 3 & 0 \\ 0 & 3 \end{vmatrix} = ?$ e. $\begin{vmatrix} 0 & 3 \\ 3 & 0 \end{vmatrix} = ?$ f. $\begin{vmatrix} 0 & 3 \\ 0 & 3 \end{vmatrix} = ?$

BIBLIOGRAPHICAL NOTE

Matrices and linear algebra are important mathematical fields in terms of applications to more advanced study in chemical kinetics, and also more generally in terms of usefulness throughout physical chemistry and chemical physics. Quantum theory, especially as applied to molecular orbitals, molecular vibrations, and molecular spectra, makes extensive use of matrices and matrix equations. The symmetry properties of molecules and crystals are usually discussed in terms of group theory, a theory in which matrices play a central role. An excellent first book about matrix theory and linear algebra, presupposing modest secondary school mathematics preparation, is P. J. Davis, *The Mathematics of Matrices*, Waltham, Mass.: Blaisdell Publishing Company, 1965.

Two paperback books written for a young British audience are G. Matthews, *Matrices 1* and *Matrices 2*, Contemporary School Mathematics Series, Boston: Houghton Mifflin Company, 1964. *Matrices 1* introduces matrices via codes, using matrices as message encoders and decoders. In *Matrices 2*, the subject of transformation matrices is discussed at first in terms of the Cinderella story.

Teaching of some of the concepts of matrix theory at the preschool and primary school level is one of the purposes of the game *Attributes*, available from the Webster Division of McGraw-Hill, Inc. The *Attributes* game materials can be challenging for any age group and provide an enjoyable introduction to matrices for the teacher as well as the student.

Appendix **II**
The Exponential Function and
Its Derivative

Solutions of many of the differential equations in this book are written in terms of the descending exponential function, e^{-x}, where e is the number given by

$$e = \lim_{n \to \infty} \left(1 + \frac{1}{n} \right)^n = 2.718\ 281\ 828 \cdots$$

When e^{-x} is plotted versus x, from $x = 0$ toward infinity, the resulting curve starts at

$$e^{-x} = 1.000 \cdots \qquad x = 0$$

The value of the function decreases continuously and monotonically, and as x increases without upper bound,

$$\lim_{x \to \infty} e^{-x} = 0$$

The function e^{-x} can be evaluated, for a particular value of x, to any desired number of significant decimal places by summing a sufficient number of terms in the infinite series

$$e^{-x} = 1 - x + \frac{x^2}{2!} - \frac{x^3}{3!} + \cdots + \frac{(-1)^n x^n}{n!} + \cdots$$

TABLE II.1.   Slope of an Exponential Curve: Two Instances

x	e^{-x}	$e^{-[x+0.0001]}$	*Slope*
0.0000	1.00000 0000	0.99990 0005	−0.99995
0.1000	0.90483 7418	0.90474 6939	−0.90479
0.2000	0.81873 0753	0.81864 8884	−0.81869
0.3000	0.74081 8221	0.74074 4143	−0.74078
0.4000	0.67032 0046	0.67025 3017	−0.67029
0.5000	0.60653 0660	0.60647 0010	−0.60650
0.6000	0.54881 1636	0.54875 6758	−0.54878
0.7000	0.49658 5304	0.49653 5648	−0.49656
0.8000	0.44932 8964	0.44928 4033	−0.44931
0.9000	0.40656 9660	0.40652 9005	−0.40655

x	e^{-10x}	$e^{-10[x+0.00001]}$	*Slope*
0.00000	1.00000 0000	0.99990 0005	−9.9995
0.01000	0.90483 7418	0.90474 6939	−9.0479
0.02000	0.81873 0753	0.81864 8884	−8.1869
0.03000	0.74081 8221	0.74074 4143	−7.4078
0.04000	0.67032 0046	0.67025 3017	−6.7029
0.05000	0.60653 0660	0.60647 0010	−6.0650
0.06000	0.54881 1636	0.54875 6758	−5.4878
0.07000	0.49658 5304	0.49653 5648	−4.9656
0.08000	0.44932 8964	0.44928 4033	−4.4931
0.09000	0.40656 9660	0.40652 9005	−4.0655

where $n!$ (read n-factorial) is the product of the first n integers. Thus $2! \equiv 1 \cdot 2 = 2$, and $3! \equiv 1 \cdot 2 \cdot 3 = 6$.

We have symbolized the slope of a plot of e^{-x} versus x by

$$\text{slope} = \frac{de^{-x}}{dx}$$

Another phrase for "the slope of a curve of the function plotted versus x" is "the derivative of the function with respect to x." The quantity

$$\frac{de^{-x}}{dx}$$

is the derivative of e^{-x} with respect to x. Two extremely important proper-

TABLE II.2. Descending Exponential Function e^{-x} Evaluated from $x = 0.000$ to $x = 0.099$

	0.000	0.001	0.002	0.003	0.004	0.005	0.006	0.007	0.008	0.009
0.00	1.00000	0.99900	0.99800	0.99700	0.99601	0.99501	0.99402	0.99302	0.99203	0.99104
0.01	0.99005	0.98906	0.98807	0.98708	0.98610	0.98511	0.98413	0.98314	0.98216	0.98118
0.02	0.98020	0.97922	0.97824	0.97726	0.97629	0.97531	0.97434	0.97336	0.97239	0.97142
0.03	0.97045	0.96948	0.96851	0.96754	0.96657	0.96561	0.96464	0.96368	0.96271	0.96175
0.04	0.96079	0.95983	0.95887	0.95791	0.95695	0.95600	0.95504	0.95409	0.95313	0.95218
0.05	0.95123	0.95028	0.94933	0.94838	0.94743	0.94649	0.94554	0.94459	0.94365	0.94271
0.06	0.94176	0.94082	0.93988	0.93894	0.93800	0.93707	0.93613	0.93520	0.93426	0.93333
0.07	0.93239	0.93146	0.93053	0.92960	0.92867	0.92774	0.92682	0.92589	0.92496	0.92404
0.08	0.92312	0.92219	0.92127	0.92035	0.91943	0.91851	0.91759	0.91668	0.91576	0.91485
0.09	0.91393	0.91302	0.91211	0.91119	0.91028	0.90937	0.90846	0.90756	0.90665	0.90574

TABLE II.3. Descending Exponential Function e^{-x} Evaluated from $x = 0.00$ to $x = 0.99$

	0.00	0.01	0.02	0.03	0.04	0.05	0.06	0.07	0.08	0.09
0.0	1.00000	0.99005	0.98020	0.97045	0.96079	0.95123	0.94176	0.93239	0.92312	0.91393
0.1	0.90484	0.89583	0.88692	0.87810	0.86936	0.86071	0.85214	0.84366	0.83527	0.82696
0.2	0.81873	0.81058	0.80252	0.79453	0.78663	0.77880	0.77105	0.76338	0.75578	0.74826
0.3	0.74082	0.73345	0.72615	0.71892	0.71177	0.70469	0.69768	0.69073	0.68386	0.67706
0.4	0.67032	0.66365	0.65705	0.65051	0.64404	0.63763	0.63128	0.62500	0.61878	0.61263
0.5	0.60653	0.60050	0.59452	0.58860	0.58275	0.57695	0.57121	0.56553	0.55990	0.55433
0.6	0.54881	0.54335	0.53794	0.53259	0.52729	0.52205	0.51685	0.51171	0.50662	0.50158
0.7	0.49659	0.49164	0.48675	0.48191	0.47711	0.47237	0.46767	0.46301	0.45841	0.45384
0.8	0.44933	0.44486	0.44043	0.43605	0.43171	0.42741	0.42316	0.41895	0.41478	0.41066
0.9	0.40657	0.40252	0.39852	0.39455	0.39063	0.38674	0.38289	0.37908	0.37531	0.37158

TABLE II.4. Descending Exponential Function e^{-x} Evaluated from $x = 0.0$ to $x = 9.9$

	0.0	0.1	0.2	0.3	0.4	0.5	0.6	0.7	0.8	0.9
0.	1.00000	0.90484	0.81873	0.74082	0.67032	0.60653	0.54881	0.49659	0.44933	0.40657
1.	0.36788	0.33287	0.30119	0.27253	0.24660	0.22313	0.20190	0.18268	0.16530	0.14957
2.	0.13534	0.12246	0.11080	0.10026	0.09072	0.08208	0.07427	0.06721	0.06081	0.05502
3.	0.04979	0.04505	0.04076	0.03688	0.03337	0.03020	0.02732	0.02472	0.02237	0.02024
4.	0.01832	0.01657	0.01500	0.01357	0.01228	0.01111	0.01005	0.00910	0.00823	0.00745
5.	0.00674	0.00610	0.00552	0.00499	0.00452	0.00409	0.00370	0.00335	0.00303	0.00274
6.	0.00248	0.00224	0.00203	0.00184	0.00166	0.00150	0.00136	0.00123	0.00111	0.00101
7.	0.00091	0.00083	0.00075	0.00068	0.00061	0.00055	0.00050	0.00045	0.00041	0.00037
8.	0.00034	0.00030	0.00027	0.00025	0.00022	0.00020	0.00018	0.00017	0.00015	0.00014
9.	0.00012	0.00011	0.00010	0.00009	0.00008	0.00007	0.00007	0.00006	0.00006	0.00005

ties of the exponential function can be written in terms of the derivatives:

$$\frac{de^{-x}}{dx} = -e^{-x} \tag{II.1}$$

$$\frac{de^{-mx}}{dx} = -me^{-mx} \tag{II.2}$$

The descending exponential function is the negative of its own derivative. The derivative of e^{-mx} with respect to x is proportional to e^{-mx}, the proportionality factor being $-m$. These two properties are illustrated in Table II.1. In the first tabulation, the value of e^{-x} is recorded at several values of x, and also at values of $[x + 0.0001]$. Each pair e^{-x} and $e^{-[x+0.0001]}$ is used to calculate the slope of the curve, a slope evaluated at $e^{-[x+0.00005]}$. The slope of the curve is seen to be the same as the function itself, except for the change of sign. In the second tabulation, the value of e^{-10x} is tabulated, and the slope is found to be $-10e^{-10x}$, just as required by Equation (II.2).

Tables II.2, II.3, and II.4 give values of the descending exponential function for use in making calculations. These tables can be extended by taking advantage of the relationship

$$e^x e^z = e^{x+z}$$

For more extensive tabulations, reference is made to M. Abramowitz and I. A. Stegun, eds., *Handbook of Mathematical Functions with Formulas, Graphs, and Mathematical Tables*, National Bureau of Standards Applied Mathematics Series No. 55, Washington, D.C.: U.S. Government Printing Office, 1968. Table 4.4 gives the exponential function with 18 significant decimal places.

Appendix III

Logarithms to the Base e

Associated with the exponential function is the logarithmic function to the base e. We write $\ln x$ to mean "the logarithm to the base e of x." A fundamental relationship, defining the logarithmic function and showing the intimate connection between the logarithmic and exponential functions, is

$$\ln [e^{-x}] = -x$$

Some other logarithmic identities that are useful in making calculations include

$$\ln [xz] = \ln x + \ln z$$

$$\ln \left(\frac{x}{z}\right) = \ln x - \ln z$$

The numerical values of $\ln x$ given in Tables III.1 and III.2 cover a sufficiently wide range of x values to be useful in constructing plots of the type

$$\ln | A_t - A_{t+\tau} | \text{ versus } t$$
$$\ln | A_t - A_\infty | \text{ versus } t$$

Note that the *slope* of such plots is not affected by multiplication of each value of $| A_t - A_{t+\tau} |$ or of $| A_t - A_\infty |$ by a constant factor. The chemist

TABLE III.1. Logarithmic Function to Base e, ln x, Evaluated from $x = 0.000$ to $x = 0.179$

	0.000	0.001	0.002	0.003	0.004	0.005	0.006	0.007	0.008	0.009
0.00	$-\infty$	-6.9078	-6.2146	-5.8091	-5.5215	-5.2983	-5.1160	-4.9618	-4.8283	-4.7105
0.01	-4.6052	-4.5099	-4.4228	-4.3428	-4.2687	-4.1997	-4.1352	-4.0745	-4.0174	-3.9633
0.02	-3.9120	-3.8632	-3.8167	-3.7723	-3.7297	-3.6889	-3.6497	-3.6119	-3.5756	-3.5405
0.03	-3.5066	-3.4738	-3.4420	-3.4112	-3.3814	-3.3524	-3.3242	-3.2968	-3.2702	-3.2442
0.04	-3.2189	-3.1942	-3.1701	-3.1466	-3.1236	-3.1011	-3.0791	-3.0576	-3.0366	-3.0159
0.05	-2.9957	-2.9759	-2.9565	-2.9375	-2.9188	-2.9004	-2.8824	-2.8647	-2.8473	-2.8302
0.06	-2.8134	-2.7969	-2.7806	-2.7646	-2.7489	-2.7334	-2.7181	-2.7031	-2.6882	-2.6736
0.07	-2.6593	-2.6451	-2.6311	-2.6173	-2.6037	-2.5903	-2.5770	-2.5639	-2.5510	-2.5383
0.08	-2.5257	-2.5133	-2.5010	-2.4889	-2.4769	-2.4651	-2.4534	-2.4418	-2.4304	-2.4191
0.09	-2.4079	-2.3969	-2.3860	-2.3752	-2.3645	-2.3539	-2.3434	-2.3330	-2.3228	-2.3126
0.10	-2.3026	-2.2926	-2.2828	-2.2730	-2.2634	-2.2538	-2.2443	-2.2349	-2.2256	-2.2164
0.11	-2.2073	-2.1982	-2.1893	-2.1804	-2.1716	-2.1628	-2.1542	-2.1456	-2.1371	-2.1286
0.12	-2.1203	-2.1120	-2.1037	-2.0956	-2.0875	-2.0794	-2.0715	-2.0636	-2.0557	-2.0479
0.13	-2.0402	-2.0326	-2.0250	-2.0174	-2.0099	-2.0025	-1.9951	-1.9878	-1.9805	-1.9733
0.14	-1.9661	-1.9590	-1.9519	-1.9449	-1.9379	-1.9310	-1.9241	-1.9173	-1.9105	-1.9038
0.15	-1.8971	-1.8905	-1.8839	-1.8773	-1.8708	-1.8643	-1.8579	-1.8515	-1.8452	-1.8389
0.16	-1.8326	-1.8264	-1.8202	-1.8140	-1.8079	-1.8018	-1.7958	-1.7898	-1.7838	-1.7779
0.17	-1.7720	-1.7661	-1.7603	-1.7545	-1.7487	-1.7430	-1.7373	-1.7316	-1.7260	-1.7204

TABLE III.2. Logarithmic Function to Base e, ln x, Evaluated from $x = 0.00$ to $x = 1.79$

	0.00	0.01	0.02	0.03	0.04	0.05	0.06	0.07	0.08	0.09
0.0	$-\infty$	-4.6052	-3.9120	-3.5066	-3.2189	-2.9957	-2.8134	-2.6593	-2.5257	-2.4079
0.1	-2.3026	-2.2073	-2.1203	-2.0402	-1.9661	-1.8971	-1.8326	-1.7720	-1.7148	-1.6607
0.2	-1.6094	-1.5606	-1.5141	-1.4697	-1.4271	-1.3863	-1.3471	-1.3093	-1.2730	-1.2379
0.3	-1.2040	-1.1712	-1.1394	-1.1087	-1.0788	-1.0498	-1.0217	-0.9943	-0.9676	-0.9416
0.4	-0.9163	-0.8916	-0.8675	-0.8440	-0.8210	-0.7985	-0.7765	-0.7550	-0.7340	-0.7133
0.5	-0.6931	-0.6733	-0.6539	-0.6349	-0.6162	-0.5978	-0.5798	-0.5621	-0.5447	-0.5276
0.6	-0.5108	-0.4943	-0.4780	-0.4620	-0.4463	-0.4308	-0.4155	-0.4005	-0.3857	-0.3711
0.7	-0.3567	-0.3425	-0.3285	-0.3147	-0.3011	-0.2877	-0.2744	-0.2614	-0.2485	-0.2357
0.8	-0.2231	-0.2107	-0.1985	-0.1863	-0.1744	-0.1625	-0.1508	-0.1393	-0.1278	-0.1165
0.9	-0.1054	-0.0943	-0.0834	-0.0726	-0.0619	-0.0513	-0.0408	-0.0305	-0.0202	-0.0101
1.0	0.0000	0.0100	0.0198	0.0296	0.0392	0.0488	0.0583	0.0677	0.0770	0.0862
1.1	0.0953	0.1044	0.1133	0.1222	0.1310	0.1398	0.1484	0.1570	0.1655	0.1740
1.2	0.1823	0.1906	0.1989	0.2070	0.2151	0.2231	0.2311	0.2390	0.2469	0.2546
1.3	0.2624	0.2700	0.2776	0.2852	0.2927	0.3001	0.3075	0.3148	0.3221	0.3293
1.4	0.3365	0.3436	0.3507	0.3577	0.3646	0.3716	0.3784	0.3853	0.3920	0.3988
1.5	0.4055	0.4121	0.4187	0.4253	0.4318	0.4383	0.4447	0.4511	0.4574	0.4637
1.6	0.4700	0.4762	0.4824	0.4886	0.4947	0.5008	0.5068	0.5128	0.5188	0.5247
1.7	0.5306	0.5365	0.5423	0.5481	0.5539	0.5596	0.5653	0.5710	0.5766	0.5822

should, therefore, feel free to move the decimal point in the entire series of data readings so as to facilitate use of the logarithm tables.

More extensive tabulations of logarithms to the base e are found in M. Abramowitz and I. A. Stegun, eds., *Handbook of Mathematical Functions with Formulas, Graphs, and Mathematical Tables*, National Bureau of Standards Applied Mathematics Series No. 55, Washington, D.C.: U.S. Government Printing Office, 1968. Table 4.2 gives logarithms to the base e with 16 decimal places.

Appendix **IV**

Logarithms to the Base 10*

* From *Rinehart Mathematical Tables, Formulas and Curves* by Harold D. Larsen. Copyright 1948, 1953 by Harold D. Larsen. Reprinted by permission of Holt, Rinehart and Winston, Inc.

N	0	1	2	3	4	5	6	7	8	9
10	0000	0043	0086	0128	0170	0212	0253	0294	0334	0374
11	0414	0453	0492	0531	0569	0607	0645	0682	0719	0755
12	0792	0828	0864	0899	0934	0969	1004	1038	1072	1106
13	1139	1173	1206	1239	1271	1303	1335	1367	1399	1430
14	1461	1492	1523	1553	1584	1614	1644	1673	1703	1732
15	1761	1790	1818	1847	1875	1903	1931	1959	1987	2014
16	2041	2068	2095	2122	2148	2175	2201	2227	2253	2279
17	2304	2330	2355	2380	2405	2430	2455	2480	2504	2529
18	2553	2577	2601	2625	2648	2672	2695	2718	2742	2765
19	2788	2810	2833	2856	2878	2900	2923	2945	2967	2989
20	3010	3032	3054	3075	3096	3118	3139	3160	3181	3201
21	3222	3243	3263	3284	3304	3324	3345	3365	3385	3404
22	3424	3444	3464	3483	3502	3522	3541	3560	3579	3598
23	3617	3636	3655	3674	3692	3711	3729	3747	3766	3784
24	3802	3820	3838	3856	3874	3892	3909	3927	3945	3962
25	3979	3997	4014	4031	4048	4065	4082	4099	4116	4133
26	4150	4166	4183	4200	4216	4232	4249	4265	4281	4298
27	4314	4330	4346	4362	4378	4393	4409	4425	4440	4456
28	4472	4487	4502	4518	4533	4548	4564	4579	4594	4609
29	4624	4639	4654	4669	4683	4698	4713	4728	4742	4757
30	4771	4786	4800	4814	4829	4843	4857	4871	4886	4900
31	4914	4928	4942	4955	4969	4983	4997	5011	5024	5038
32	5051	5065	5079	5092	5105	5119	5132	5145	5159	5172
33	5185	5198	5211	5224	5237	5250	5263	5276	5289	5302
34	5315	5328	5340	5353	5366	5378	5391	5403	5416	5428
35	5441	5453	5465	5478	5490	5502	5514	5527	5539	5551
36	5563	5575	5587	5599	5611	5623	5635	5647	5658	5670
37	5682	5694	5705	5717	5729	5740	5752	5763	5775	5786
38	5798	5809	5821	5832	5843	5855	5866	5877	5888	5899
39	5911	5922	5933	5944	5955	5966	5977	5988	5999	6010
40	6021	6031	6042	6053	6064	6075	6085	6096	6107	6117
41	6128	6138	6149	6160	6170	6180	6191	6201	6212	6222
42	6232	6243	6253	6263	6274	6284	6294	6304	6314	6325
43	6335	6345	6355	6365	6375	6385	6395	6405	6415	6425
44	6435	6444	6454	6464	6474	6484	6493	6503	6513	6522
45	6532	6542	6551	6561	6571	6580	6590	6599	6609	6618
46	6628	6637	6646	6656	6665	6675	6684	6693	6702	6712
47	6721	6730	6739	6749	6758	6767	6776	6785	6794	6803
48	6812	6821	6830	6839	6848	6857	6866	6875	6884	6893
49	6902	6911	6920	6928	6937	6946	6955	6964	6972	6981
50	6990	6998	7007	7016	7024	7033	7042	7050	7059	7067
51	7076	7084	7093	7101	7110	7118	7126	7135	7143	7152
52	7160	7168	7177	7185	7193	7202	7210	7218	7226	7235
53	7243	7251	7259	7267	7275	7284	7292	7300	7308	7316
54	7324	7332	7340	7348	7356	7364	7372	7380	7388	7396
N	0	1	2	3	4	5	6	7	8	9

10.0—Four-Place Common Logarithms of Numbers—54.9

N	0	1	2	3	4	5	6	7	8	9
55	7404	7412	7419	7427	7435	7443	7451	7459	7466	7474
56	7482	7490	7497	7505	7513	7520	7528	7536	7543	7551
57	7559	7566	7574	7582	7589	7597	7604	7612	7619	7627
58	7634	7642	7649	7657	7664	7672	7679	7686	7694	7701
59	7709	7716	7723	7731	7738	7745	7752	7760	7767	7774
60	7782	7789	7796	7803	7810	7818	7825	7832	7839	7846
61	7853	7860	7868	7875	7882	7889	7896	7903	7910	7917
62	7924	7931	7938	7945	7952	7959	7966	7973	7980	7987
63	7993	8000	8007	8014	8021	8028	8035	8041	8048	8055
64	8062	8069	8075	8082	8089	8096	8102	8109	8116	8122
65	8129	8136	8142	8149	8156	8162	8169	8176	8182	8189
66	8195	8202	8209	8215	8222	8228	8235	8241	8248	8254
67	8261	8267	8274	8280	8287	8293	8299	8306	8312	8319
68	8325	8331	8338	8344	8351	8357	8363	8370	8376	8382
69	8388	8395	8401	8407	8414	8420	8426	8432	8439	8445
70	8451	8457	8463	8470	8476	8482	8488	8494	8500	8506
71	8513	8519	8525	8531	8537	8543	8549	8555	8561	8567
72	8573	8579	8585	8591	8597	8603	8609	8615	8621	8627
73	8633	8639	8645	8651	8657	8663	8669	8675	8681	8686
74	8692	8698	8704	8710	8716	8722	8727	8733	8739	8745
75	8751	8756	8762	8768	8774	8779	8785	8791	8797	8802
76	8808	8814	8820	8825	8831	8837	8842	8848	8854	8859
77	8865	8871	8876	8882	8887	8893	8899	8904	8910	8915
78	8921	8927	8932	8938	8943	8949	8954	8960	8965	8971
79	8976	8982	8987	8993	8998	9004	9009	9015	9020	9025
80	9031	9036	9042	9047	9053	9058	9063	9069	9074	9079
81	9085	9090	9096	9101	9106	9112	9117	9122	9128	9133
82	9138	9143	9149	9154	9159	9165	9170	9175	9180	9186
83	9191	9196	9201	9206	9212	9217	9222	9227	9232	9238
84	9243	9248	9253	9258	9263	9269	9274	9279	9284	9289
85	9294	9299	9304	9309	9315	9320	9325	9330	9335	9340
86	9345	9350	9355	9360	9365	9370	9375	9380	9385	9390
87	9395	9400	9405	9410	9415	9420	9425	9430	9435	9440
88	9445	9450	9455	9460	9465	9469	9474	9479	9484	9489
89	9494	9499	9504	9509	9513	9518	9523	9528	9533	9538
90	9542	9547	9552	9557	9562	9566	9571	9576	9581	9586
91	9590	9595	9600	9605	9609	9614	9619	9624	9628	9633
92	9638	9643	9647	9652	9657	9661	9666	9671	9675	9680
93	9685	9689	9694	9699	9703	9708	9713	9717	9722	9727
94	9731	9736	9741	9745	9750	9754	9759	9763	9768	9773
95	9777	9782	9786	9791	9795	9800	9805	9809	9814	9818
96	9823	9827	9832	9836	9841	9845	9850	9854	9859	9863
97	9868	9872	9877	9881	9886	9890	9894	9899	9903	9908
98	9912	9917	9921	9926	9930	9934	9939	9943	9948	9952
99	9956	9961	9965	9969	9974	9978	9983	9987	9991	9996
N	**0**	**1**	**2**	**3**	**4**	**5**	**6**	**7**	**8**	**9**

55.0—Four-Place Common Logarithms of Numbers—99.9

Appendix **V**
Additional Topics in Chemical Dynamics:
A Selective Bibliography for Further Reading and Intensive Study

Amdur, I., and G. G. Hammes, *Chemical Kinetics: Principles and Selected Topics*, New York: McGraw-Hill, Inc., 1966.

Ames, W. F., *Nonlinear Ordinary Differential Equations in Transport Processes*, New York: Academic Press, Inc., 1968. Chapter 3 begins with a section on "Matrices and Chemical Reactions."

Ashmore, P. G., *Catalysis and Inhibition of Chemical Reactions*, London: Butterworths, 1963.

Benson, S. W., *The Foundations of Chemical Kinetics*, New York: McGraw-Hill, Inc., 1960.

Bradley, J. N., *Chemical Applications of the Shock Tube*, London: The Royal Institute of Chemistry, 1963 (1963 Lecture Series, No. 6).

Bray, H. G., and K. White, *Kinetics and Thermodynamics in Biochemistry*, New York: Academic Press, Inc., 2nd ed., 1966. See especially Chap. 8, "Introduction to the Kinetics of Tracers in Reaction Systems."

Calvert, J. G., and J. N. Pitts, Jr., *Photochemistry*, New York: John Wiley & Sons, Inc., 1966.

Dainton, F. S., *Chain Reactions: An Introduction*, New York: John Wiley & Sons, Inc., 2nd ed., 1966.

Discussions of the Faraday Society, No. 17, "The Study of Fast Reactions," 1954; No. 39, "The Kinetics of Proton Transfer Processes," 1965; No. 44, "Molecular Dynamics of the Chemical Reactions of Gases," 1967.

Frost, A. A., and R. G. Pearson, *Kinetics and Mechanism: A Study of Homogeneous Chemical Reactions*, New York: John Wiley & Sons, Inc., 2nd ed., 1961.

Gavalas, G. R., *Nonlinear Differential Equations of Chemically Reacting Systems*, New York: Springer-Verlag, Inc., 1968.

Jencks, W. P., *Catalysis in Chemistry and Enzymology*, New York: McGraw-Hill, Inc., 1969.

Johnston, H. S., *Gas Phase Reaction Rate Theory*, New York: The Ronald Press Company, 1966.

Jones, M. M., *Ligand Reactivity and Catalysis*, New York: Academic Press, Inc., 1968.

Laidler, K. J., *The Chemical Kinetics of Excited States*, New York: Oxford University Press, 1955.

Mechanisms of Inorganic Reactions (a collection of symposium papers), Advances in Chemistry Series No. 49, Washington, D.C.: American Chemical Society, 1965.

Nikitin, E. E., *Theory of Thermally Induced Gas Phase Reactions*, Bloomington, Ind.: Indiana University Press, 1966. Translated from Russian by E. W. Schlag.

Reactions of Coordinated Ligands and Homogeneous Catalysis (a collection of symposium papers), Advances in Chemistry Series No. 37, Washington, D.C.: American Chemical Society, 1963.

Redistribution Reactions in Chemistry, New York: The New York Academy of Sciences, *Annals*, Vol. 159, Art. 1, pp. 1–334, 1969.

Rescigno, A., and G. Segre, *Drug and Tracer Kinetics*, Waltham, Mass.: Blaisdell Publishing Company, 1966. Translated from Italian by P. Ariotti.

Ross, J., ed., "Molecular Beams," vol. 10 of *Advances in Chemical Physics*, New York: John Wiley & Sons, Inc. (Interscience Division), 1966.

Slater, N. B., *Theory of Unimolecular Reactions*, Ithaca, N.Y.: Cornell University Press, 1959.

Sykes, A. G., "Further Advances in the Study of Mechanisms of Redox Reactions," in *Advances in Inorganic Chemistry and Radiochemistry*, vol. 10, New York: Academic Press, Inc., 1967, pp. 153–245.

Wells, P. R., *Linear Free Energy Relationships*, New York: Academic Press, Inc., 1968.

Westley, J., *Enzymic Catalysis*, New York: Harper & Row, Inc., 1969.

Appendix **VI**
Greek Alphabet

alpha	A	α		nu	N	ν
beta	B	β		xi	Ξ	ξ
gamma	Γ	γ		omicron	O	o
delta	Δ	δ		pi	Π	π
epsilon	E	ϵ		rho	P	ρ
zeta	Z	ζ		sigma	Σ	σ
eta	H	η		tau	T	τ
theta	Θ	θ		upsilon	Υ	υ
iota	I	ι		phi	Φ	ϕ
kappa	K	κ		chi	X	χ
lambda	Λ	λ		psi	Ψ	ψ
mu	M	μ		omega	Ω	ω

Appendix **VII**

Answers to Problems

CHAPTER 2

2.1. $T = 1.00 \rightarrow A = 0.000 \pm 0.002$
$T = 0.10 \rightarrow A = 1.00 \pm 0.02$
$T = 0.01 \rightarrow A = 2.00 \left(\begin{array}{c} +0.3 \\ -0.18 \end{array}\right.$
$T = 0.97 \rightarrow A = 0.013 \pm 0.002$
$T = 0.96 \rightarrow A = 0.018 \pm 0.002$
$T = 0.03 \rightarrow A = 1.52 \pm 0.07$
$T = 0.04 \rightarrow A = 1.40 \pm 0.05$

2.2. $pH = 14.00 \rightarrow (H^+) = 10^{-14}$ mole/liter
$pH = 7.00 \rightarrow (H^+) = 10^{-7}$ mole/liter
$pH = 2.00 \rightarrow (H^+) = 10^{-2}$ mole/liter
$pH = 2.01 \rightarrow (H^+) = 10^{-2.01} = 10^{[-3+.99]} = 9.77 \times 10^{-3}$ mole/liter
$pH = 1.99 \rightarrow (H^+) = 10^{-1.99} = 10^{[-2+.01]} = 1.02 \times 10^{-2}$ mole/liter
$pH = 2.50 \rightarrow (H^+) = 10^{-2.50} = 10^{[-3+.50]} = 3.16 \times 10^{-3}$ mole/liter
Throughout this problem it was assumed that we can accept the approximation: $pH = -\log_{10}(H^+)$.

2.3. $l = 10.0$ mm $\rightarrow A = 2.0$
$l = 1.00$ mm $\rightarrow A = 0.20$
$l = .10$ mm $\rightarrow A = 0.020$

CHAPTER 3

3.1. The coefficient c can take on the two values c_0 and c_1. Equation (3.31) shows that $c_0 = (\overline{C})$, and because $(\overline{C})$ must be greater than zero, c_0 cannot be equal to zero. Equation (3.29) imposes the requirement that $c_0 + c_1$ be equal to the initial concentration of (C). Only if the initial concentration of C is equal to the equilibrium concentration of C can c_1 be equal to zero. Such a situation would be called a trivial case, because all concentrations would have their equilibrium values at time zero, and there would be no rates of reaction to observe.

3.2. Let us substitute into differential equation (3.7). The derivative $d(C)/dt$ is obtained by differentiating Equation (3.27) with respect to t. (The derivative of a sum of terms is the sum of the derivatives of the individual terms. The derivative of a constant is zero. The derivative of $c_1 e^{-m_1 t}$ with respect to t is $-m_1 c_1 e^{-m_1 t}$.) We get from (3.27)

$$\frac{d(C)}{dt} = -m_1 c_1 e^{-m_1 t}$$

and proceed to substitute this derivative, Equation (3.26), and Equation (3.27) directly into Equation (3.7):

$$-m_1 c_1 e^{-m_1 t} = k_1 b_0 + k_1 b_1 e^{-m_1 t} - k_{-1} c_0 - k_{-1} c_1 e^{-m_1 t} \qquad \text{(I)}$$

From Equation (3.7) evaluated at equilibrium we know that

$$k_1(\overline{B}) - k_{-1}(\overline{C}) = 0$$

From Equations (3.26) and (3.27) evaluated at equilibrium we find that

$$(\overline{B}) = b_0 \qquad (\overline{C}) = c_0$$

It follows that

$$k_1 b_0 - k_{-1} c_0 = 0$$

Two terms thus disappear from Equation (I), leaving

$$-m_1 c_1 e^{-m_1 t} = k_1 b_1 e^{-m_1 t} - k_{-1} c_1 e^{-m_1 t}$$

Division by the factor $e^{-m_1 t}$ gives

$$-m_1 c_1 = k_1 b_1 - k_{-1} c_1 \qquad \text{(II)}$$

which is a statement of the necessary relationships among the time-invariant quantities required so that Equations (3.26) and (3.27) can be valid solutions of Equation (3.7). Substitution into differential equation (3.8) proceeds in exactly the same way.

3.3. $\tau = 1/m_1$.

3.4. The straight-line plot results because $\log_{10} x$ is proportional to $\ln x$. This fact can be verified by comparison of the logarithm tables in Appendix III and Appendix IV. This proportionality can also be demonstrated as follows. Since $\log_{10} 10 = 1$, we can write

$$y(\log_{10} e) = y(\log_{10} e)(\log_{10} 10) \qquad (I)$$

Since $y(\log_{10} a) = \log_{10} a^y$, Equation (I) can be written as

$$\log_{10}(e^y) = \log_{10}(10^{y\log_{10}e}) \qquad (II)$$

Since $\log_{10} \phi$ is a continuous, single-valued function of the variable ϕ, Equation (II) requires that

$$e^y = 10^{y\log_{10}e}$$

Now let $x = e^y$, giving

$$\ln x = \ln(e^y) = \ln(10^{y\log_{10}e})$$
$$= y(\log_{10} e)(\ln 10)$$
$$= (\log_{10} e^y)(\ln 10)$$
$$= (\log_{10} x)(\ln 10) \qquad (III)$$

We take advantage of a table of logarithms to find that

$$\ln 10 = 2.302585\cdots$$

Equation (III) thus becomes

$$\ln x = 2.303 \log_{10} x$$

3.5. *Hint*: Start with an equation such as (3.47). For the state $1/n$th of the way to equilibrium, $[(B)_t - (\overline{B})]$ will have changed to $[1 - (1/n)][(B)_t - (\overline{B})]$.

3.6. $m_1 = 0.0050$ second^{-1}.

3.7. If $t = \tau$ initially, then Equations (3.28) and (3.29) become

$$(B)_{time=\tau} = b_0 + b_1 e^{-m_1 \tau}$$

$$(C)_{time=\tau} = c_0 + c_1 e^{-m_1 \tau}$$

At all later times, t units after τ, time is given by $\tau + t$, and the instantaneous concentrations are

$$(B) = b_0 + b_1 e^{-m_1 (\tau+t)} = (\overline{B}) + \{b_1 e^{-m_1 \tau}\} e^{-m_1 t}$$

$$(C) = c_0 + c_1 e^{-m_1 (\tau+t)} = (\overline{C}) + \{c_1 e^{-m_1 \tau}\} e^{-m_1 t}$$

These are equivalent to Equations (3.32) and (3.33), since both sets of equations are equivalent to

$$(B) - (\overline{B}) = \{constant\} e^{-m_1 t}$$

$$(C) - (\overline{C}) = \{constant\} e^{-m_1 t}$$

CHAPTER 5

5.1. For $k_1 \gg k_2$:

$$(B) = (A)_0 \left(\frac{k_{-2}}{k_2} - e^{-k_1 t} + e^{-k_2 t} \right)$$

The concentration of B passes through a maximum, at which time virtually all the original A is present as the intermediate B.
For $k_2 \gg k_1$:

$$(B) = (A)_0 \left(\frac{k_{-2}}{k_2} - \frac{k_1}{k_2} \left[e^{-k_2 t} - e^{-k_1 t} \right) \right)$$

The value of (B) passes through a maximum, but the concentration of B is always very small compared to either (A) or (C).

5.2. It appears that under all circumstances, $k_{-2} < m_1$. Substitution of the values $k_1 = k_{-1} = k_2 = 1$, $k_{-2} = 2$ into Equations (5.29) and (5.30) yields the result $m_1 = 3.6$, $m_2 = 1.4$. If we try $k_1 = k_{-1} = k_2 = 1$, $k_{-2} = 1000$, we get $m_1 = 1001.001$, $m_2 = 1.999$.

5.3. The second derivative of (C) with respect to time is equal to

$$\frac{(A)_0 k_1 k_2}{m_1 - m_2} \{ m_1 e^{-m_1 t} - m_2 e^{-m_2 t} \}$$

For the second derivative to be zero, it is necessary that

$$m_1 e^{-m_1 t} = m_2 e^{-m_2 t}$$

and there is always a value of t which satisfies this equation. Hence there is always an inflection in the (C) versus t curve.

The second derivative of (B) with respect to time is

$$\frac{(A)_0 k_1}{m_1 - m_2} \{ -m_1 [m_1 - k_{-2}] e^{-m_1 t} + m_2 [m_2 - k_{-2}] e^{-m_2 t} \}$$

For the second derivative to be zero

$$m_1 [m_1 - k_{-2}] e^{-m_1 t} = m_2 [m_2 - k_{-2}] e^{-m_2 t}$$

For this equation to be valid, $[m_1 - k_{-2}]$ and $[m_2 - k_{-2}]$ must have the same sign.

5.4.
$$m_1 + m_2 + m_3 = k_1 + k_{-1} + k_2 + k_{-2} + k_3 + k_{-3}$$

$$m_1 m_2 + m_1 m_3 + m_2 m_3 = k_1 k_2 + k_1 k_{-2} + k_1 k_3 + k_1 k_{-3} + k_{-1} k_3$$
$$+ k_2 k_3 + k_{-1} k_{-3} + k_2 k_{-3} + k_{-1} k_{-2}$$

$$m_1 m_2 m_3 = k_1 k_2 k_3 + k_1 k_2 k_{-3} + k_1 k_{-2} k_{-3} + k_{-1} k_{-2} k_{-3}$$

5.5.
$$m_1 + m_2 + m_3 = k_1 + k_{-1} + k_2 + k_{-2} + k_3 + k_{-3}$$

$$m_1 m_2 + m_1 m_3 + m_2 m_3 = k_{-1} k_{-2} + k_{-1} k_{-3} + k_1 k_{-3} + k_{-3} k_2 + k_1 k_{-2}$$
$$+ k_3 k_{-2} + k_{-1} k_2 + k_{-1} k_3 + k_{-2} k_{-3}$$

$$m_1 m_2 m_3 = k_{-1} k_{-2} k_{-3} + k_1 k_{-2} k_{-3} + k_{-1} k_2 k_{-3} + k_{-1} k_{-2} k_3$$

5.6. The first relaxation can be analyzed by the Guggenheim method, using τ equal to about 30 minutes. The values of $\alpha_{t+\tau}$ must be estimated, and for this purpose it is helpful to make an α versus t plot of the data. The macroscopic rate constant m_1 is about 0.02 minute^{-1}. The slower relaxation can be analyzed by the infinite-time method, giving the value $m_2 = 0.0042$ minute^{-1}. The two relaxations overlap, and more reliable values of the m's can be obtained by subtracting the quantity {constant} $e^{-m_2 t}$, evaluated at the appropriate values of t, from the early-time data before making the Guggenheim plot.

5.7. 1 hour $= 3.6 \times 10^3$ seconds.
1 day $= 8.6 \times 10^4$ seconds.
1 year $= 3.2 \times 10^7$ seconds.
1 century $= 3.2 \times 10^9$ seconds.

5.8. If the m's have quite different numerical values, time intervals can be found in which Equation (5.76) can be approximated by Equation (5.78) or (5.83). Either equation, written at times t and $t + \tau$, yields equations of the same form as (3.44) and (3.45); the derivation then proceeds in the same way as the derivation which gave Equation (3.46).

CHAPTER 6

6.1. Substitute Equation (6.31) in the form

$$k_1 = \frac{k_2 k_{-1}}{k_{-2}}$$

into Equation (6.28). The sequence of operations is

$$\frac{(\bar{N})}{(\bar{M})} = \frac{k_1 + k_2(\bar{L})}{k_{-1} + k_{-2}(\bar{L})}$$

$$= \frac{k_2 k_{-1}/k_{-2} + k_2(\bar{L})}{k_{-1} + k_{-2}(\bar{L})}$$

$$= \frac{k_2 k_{-1} + k_2 k_{-2}(\bar{L})}{k_{-2} k_{-1} + k_{-2}{}^2(\bar{L})}$$

$$= \frac{k_2}{k_{-2}} \left(\frac{k_{-1} + k_{-2}(\bar{L})}{k_{-1} + k_{-2}(\bar{L})} \right) = \frac{k_2}{k_{-2}}$$

CHAPTER 8

Each of the derivations yields $m_0 = 0$. Only the additional nonzero m's are included in the answers below.

8.1. $m_1 = k_1(A)_0 + k_{-1}(C)_0$

8.2. $m_1 = k_1(\bar{A}) + k_{-1}(\bar{B})$. Note that in the absence of measurable isotope rate effect, $k_1 = k_{-1}$.

8.3. $m_1 = 4k_1(\bar{A}) + k_{-1}$. Note that

$$\frac{d(A)}{dt} = 2\{-k_1(A)^2 + k_{-1}(A_2)\}$$

since the elementary process

$$2A \rightarrow A_2$$

consumes two A molecules for each A_2 produced, and the elementary process

$$2A \leftarrow A_2$$

produces two A molecules for each A_2 consumed. Thus

$$\frac{d(A)}{dt} = -2\frac{d(A_2)}{dt}$$

8.4. $m_1 = k_1[(\overline{A}) + (\overline{B})] + k_{-1}[(\overline{C}) + (\overline{D})]$

8.5. $m_1 = k_1[(\overline{A})(\overline{B}) + (\overline{B})(\overline{C}) + (\overline{C})(\overline{A})] + k_{-1}$

8.6. $m_1 + m_2 = (A)_0[k_1 + k_2] + k_{-1} + k_{-2}$

$m_1 m_2 = (A)_0[k_1 k_{-2} + k_{-1} k_2] + k_{-1} k_{-2}$

8.7. $m_1 + m_2 = [(\overline{A}) + (\overline{B})][k_1 + k_2] + k_{-1} + k_{-2}$

$m_1 m_2 = [(\overline{A}) + (\overline{B})][k_1 k_{-2} + k_{-1} k_2] + k_{-1} k_{-2}$

8.8. $m_1 + m_2 = k_1[(\overline{A}) + (\overline{B})] + k_{-1} + k_2 + k_{-2}$

$m_1 m_2 = [(\overline{A}) + (\overline{B})][k_1 k_2 + k_1 k_{-2}] + k_{-1} k_{-2}$

8.9. $m_1 + m_2 + m_3 = (A)_0 k_1 + k_{-1} + k_2 + k_{-2} + k_3 + k_{-3}$

$m_1 m_2 + m_1 m_3 + m_2 m_3 = (A)_0 k_1[k_2 + k_{-2} + k_3 + k_{-3}]$
$+ [k_{-1} + k_2][k_3 + k_{-3}] + k_{-1} k_{-2}$

$m_1 m_2 m_3 = (A)_0 k_1[k_2 k_3 + k_2 k_{-3} + k_{-2} k_{-3}] + k_{-1} k_{-2} k_{-3}$

8.10. $m_1 + m_2 + m_3 = (A)_0 k_2 + k_1 + k_{-1} + k_{-2} + k_3 + k_{-3}$

$m_1 m_2 + m_1 m_3 + m_2 m_3 = (A)_0 k_2[k_1 + k_3 + k_{-3}]$
$+ [k_1 + k_{-1}][k_{-2} + k_3 + k_{-3}]$

$m_1 m_2 m_3 = (A)_0 k_2[k_1 k_3 + k_1 k_{-3}]$
$+ k_{-2}[k_1 k_{-3} + k_{-1} k_{-3}]$

8.11.
$$m_1 + m_2 + m_3 = (A)_0[k_1 + k_2] + k_{-1} + k_{-2} + k_3 + k_{-3}$$

$$m_1m_2 + m_1m_3 + m_2m_3 = (A)_0\{k_1[k_{-2} + k_3 + k_{-3}]$$
$$+ k_2[k + (A)_0^2\, k_1k_2k_3 + k_{-3}]\}$$
$$+ k_{-1}[k_{-2} + k_3 + k_{-3}]$$

$$m_1m_2m_3 = (A)_0k_1k_{-2}k_{-3} + (A)_0^2k_1k_2[k_3 + k_{-3}]$$
$$+ k_{-1}k_{-2}k_{-3}$$

8.12.
$$m_1 + m_2 + m_3 = (A)_0k_1 + (C)_0k_2 + (D)_0k_{-3}$$
$$+ k_{-1} + k_{-2} + k_3$$

$$m_1m_2 + m_1m_3 + m_2m_3 = (A)_0k_1\{k_{-2} + k_3\} + (C)_0k_2k_3\}$$
$$+ (D)_0k_{-3}k_{-1} + (A)_0(D)_0k_1k_{-3}$$
$$+ (C)_0(D)_0k_2k_{-3} + k_{-1}\{k_{-2} + k_3\} + (A)_0(C)_0k_1k_2$$

$$m_1m_2m_3 = (A)_0(D)_0k_1k_{-2}k_{-3} + (A)_0(C)_0k_1k_2k_3$$
$$+ (A)_0(C)_0(D)_0k_1k_2k_{-3} + (D)_0k_{-1}k_{-2}k_{-3}$$

CHAPTER 10

10.1. One numerical example is $k_{-1} = 100$, $k_2 = 100$, $k_1 = 1$, $k_{-2} = 1$. Then

$$m_1 + m_2 = 202$$
$$m_1m_2 = 201$$
$$m_1 = 200.9999$$
$$m_2 = 1.0001$$

10.2. Direct comparison of Equations (10.27) and (10.28) shows that, for the mechanism with no intermediate, $d(S)/dt = -d(P)/dt$. For Mechanism (10.38)–(10.39), the same equality holds if the intermediate X has its steady-state concentration. For this mechanism,

$$[(S)_0 + (P)_0] = (S) + (P) + (X)$$

Now, because $[(S)_0 + (P)_0]$ is a constant in time and (X) is also a constant during the steady-state phase, we can write

$$0 = \frac{d(S)}{dt} + \frac{d(P)}{dt} + 0$$

which is

$$\frac{d(S)}{dt} = -\frac{d(P)}{dt}$$

For the n-intermediate mechanism, the requirement is that all n intermediates have steady-state concentrations.

10.3. If the value of $(E)_0$ is known, two microscopic rate constants can be calculated by means of

$$\frac{V_S}{(E)_0} = k_2 \qquad \frac{V_P}{(E)_0} = k_{-1}$$

The sum $[k_2 + k_{-1}]$ can thus be calculated, and then we can evaluate

$$\frac{k_2 + k_{-1}}{K_S} = k_1 \qquad \frac{k_2 + k_{-1}}{K_P} = k_{-2}$$

10.4. The concentration (P') is zero at $t = 0$. Equation (10.133), evaluated at $t = 0$, is

$$0 = (P') = C + D$$

Thus $C = -D$.

10.5. The requirement of microscopic reversibility that equilibrium be established for each elementary reaction independently yields

$$\frac{k_1}{k_{-1}} = \frac{(\overline{X})}{(\overline{E})(\overline{S})} \qquad \text{[from Reaction (10.38)]}$$

$$\frac{k_2}{k_{-2}} = \frac{(\overline{P})(\overline{E})}{(\overline{X})} \qquad \text{[from Reaction (10.39)]}$$

The overall apparent equilibrium constant for the reaction, K', is given by

$$K' \equiv \frac{(\overline{P})}{(\overline{S})} = \frac{k_1 k_2}{k_{-1} k_{-2}}$$

The four steady-state kinetic constants defined by Equations (10.78)–(10.81) combine to give

$$\frac{V_S}{V_P} = \frac{k_2}{k_{-1}} \qquad \frac{K_P}{K_S} = \frac{k_1}{k_{-2}}$$

Finally, we can write

$$\frac{V_S K_P}{V_P K_S} = \frac{k_1 k_2}{k_{-1} k_{-2}} = K'$$

For Mechanism (10.85)–(10.87), the corresponding relationship is

$$\frac{V_S K_P}{V_P K_S} = \frac{k_1 k_2 k_3}{k_{-1} k_{-2} k_{-3}} = K'$$

10.6. *Hints:* The conservation equation for all enzyme-containing species is

$$(E)_0 = (E) + (X) + (EI)$$

Assume that the concentration of I is so much greater than the concentration of enzyme that

$$(I) \simeq (I)_0$$

Assume the steady state with respect to (X) and also (EI). For (EI), this gives

$$\frac{k_3}{k_{-3}} = \frac{(EI)}{(I)_0 (E)}$$

We can define a quantity $K_I = k_3 / k_{-3}$.

10.7. From Equation (10.83) we get

$$\lim_{(S)_0 \gg K_S} \left(\frac{d(S)}{dt} \right)_0 = - \frac{V_S (S)_0}{(S)_0} = - V_S$$

When $(S) = K_S$, Equation (10.83) becomes

$$\left(\frac{d(S)}{dt} \right)_0 = - \frac{V_S (S)_0}{2(S)_0} = - \tfrac{1}{2} V_S$$

APPENDIX I

I.1.

a. $\begin{pmatrix} 6 & 6 & 6 \\ 15 & 15 & 15 \\ 24 & 24 & 24 \end{pmatrix}$ b. $\begin{pmatrix} 1 & 2 & 3 \\ 4 & 5 & 6 \\ 7 & 8 & 9 \end{pmatrix}$

Note that the matrix

$$\begin{pmatrix} 1 & 0 & 0 \\ 0 & 1 & 0 \\ 0 & 0 & 1 \end{pmatrix}$$

has the same properties with respect to matrix multiplication as does the integer 1 in arithmetic. This matrix is often called the unitary matrix or the identity matrix.

I.2.   a. 0; b. 0; c. 1; d. 9; e. -9; f. 0.

Index